T0296005

Program Management in Defense and High Tech Environments

Best Practices and Advances in Program Management Series

Series Editor
Ginger Levin

RECENTLY PUBLISHED TITLES

Program Management in Defense and High Tech Environments
Charles Christopher McCarthy

The Self-Made Program Leader: Taking Charge in Matrix Organizations
Steve Tkalcevich

Transforming Business with Program Management: Integrating Strategy, People, Process, Technology, Structure, and Measurement
Satish P. Subramanian

Stakeholder Engagement: The Game Changer for Program Management
Amy Baugh

Making Projects Work: Effective Stakeholder and Communication Management
Lynda Bourne

Agile for Project Managers
Denise Canty

Project Planning and Project Success: The 25% Solution
Pedro Serrador

Project Health Assessment
Paul S. Royer, PMP

Portfolio Management: A Strategic Approach
Ginger Levin and John Wyzalek

Program Governance
Muhammad Ehsan Khan

Project Management for Research and Development: Guiding Innovation for Positive R&D Outcomes
Lory Mitchell Wingate

The Influential Project Manager: Winning Over Team Members and Stakeholders
Alfonso Bucero

PfMP® Exam Practice Tests and Study Guide
Ginger Levin

Program Management in Defense and High Tech Environments

Charles Christopher McCarthy

CRC Press
Taylor & Francis Group
Boca Raton   London   New York

CRC Press is an imprint of the
Taylor & Francis Group, an **informa** business
AN AUERBACH BOOK

CRC Press
Taylor & Francis Group
6000 Broken Sound Parkway NW, Suite 300
Boca Raton, FL 33487-2742

© 2016 by Taylor & Francis Group, LLC
CRC Press is an imprint of Taylor & Francis Group, an Informa business

No claim to original U.S. Government works

Printed on acid-free paper
Version Date: 20150910

International Standard Book Number-13: 978-1-4822-0838-2 (Hardback)

Visit the Taylor & Francis Web site at
http://www.taylorandfrancis.com

and the CRC Press Web site at
http://www.crcpress.com

Printed and bound by CPI Group (UK) Ltd, Croydon, CR0 4YY

This book is dedicated to my parents, wife, and children, all of whom taught me to treat everyone with respect, concern, and kindness, which I consider to be the central values of leadership. And this is also dedicated to the talented and dedicated colleagues that have enriched my professional life.

Contents

Preface xiii
List of Acronyms.. xv
About the Author.. xvii

Chapter 1 Overview: Program Management in the
Department of Defense (DoD)/High Technology
Environment.. 1

Role of the Program Manager 3
Qualifications, Experience, Talents, and Skills......................8
  Qualifications...11
  Experience 12
  Talents 13
  Skills 14
Types of Programs..15
  Research 15
  Design...16
  Production 17
Types of Contracts..18
Organizational Overview—Departmental Interfaces..........23
Summary 24
End of Chapter Questions 25
  For Discussion...25
  Written Assignments 26

Chapter 2 Learning the Ropes: Understanding the Culture,
the Customer, and the Program Capabilities................ 27

The Program in the Company Culture 27
The Program and the Customer (and His or
Her Culture) 31
  Cost...31
  Schedule 33
  Quality..33
The Program and the Team 34

End of Chapter Questions 38
    For Discussion..38
    Written Assignments 38

Chapter 3   Identifying Opportunities 39

The Program Manager's Knowledge Is Key....................39
Program Manager Opportunities...................................... 43
End of Chapter Questions 45
    For Discussion..45
    Written Assignments 46

Chapter 4   Pre-Proposal Work.. 47

Using Pre-Proposal Efforts to Develop a Winning
Proposal 51
Other Considerations..53
Summary 54
End of Chapter Questions 55
    For Discussion..55
    Written Assignments 55

Chapter 5   The Proposal Process for a "Typical Program"............. 57

The Important Pre-Proposal Period 57
To Bid or Not To Bid 58
Developing the Strategy—Getting Started 60
"Price to Win" 62
Leading the Proposal Team 63
Building the Team 64
Proposal Preparation—Storyboarding and
Team Dynamics... 66
Pricing..67
Pricing Strategies and Risk Management 68
Reviews 69
Business Reviews—The Sign-Off Process 72
Noncompetitive Proposals 73
Winning and Almost Winning the Contract—Final
Negotiations 74
Contract Refinement..75

But What If You Lose? 77
What If You Lost for the "Wrong Reason?" 78
End of Chapter Questions 79
   For Discussion 79
   Written Assignments 79
   Mini Project for a Team 80

Chapter 6 Planning the Program and Starting Work 81

The Management Part 84
The Leadership Part 87
Sourcing 92
   Engineering Labor 92
      People Brought in from Other Divisions 92
      Contract Engineers 93
      Waiting for the Right Engineers 94
      Overtime (OT) 94
      But What about Compensated OT? 95
Outsourcing Work Packages 97
In Summary 97
Outsourcing Product 98
Building the Program Culture 100
End of Chapter Questions 102
   For Discussion 102
   Written Assignments 102

Chapter 7 Running the Program 103

Leadership Styles 104
Making Progress and Monitoring Progress 108
Monitoring Progress—Metrics 111
Focusing on Quality 114
Managing the Customer 117
Identifying and Avoiding Performance Traps 119
Getting "Stuck" and Getting "Unstuck" 121
Customers as Motivators 123
Keeping Senior Management Engaged 125
   Program Reviews 126
Detecting Trouble and Determining
What to Do about It 128

When Problems Get Really Bad 131
Countervailing Forces and Priorities 133
Detecting and Avoiding "Scope Creep"—Internal 134
Detecting and Avoiding "Scope Creep"—External 138
Scope Creep—In Summary 140
Monitoring versus Controlling.. 141
Cost Control in the Trenches... 143
Monitoring Schedules—Program Reviews......................... 146
Leadership and Caring 151
Program Changes and Continuity.. 152
Managing External Changes 156
Celebrating Victories—Confronting Defeats..................... 157
Dealing with Individual Performance Problems.............. 160
Diagnosing and Resolving Problems.................................. 164
Celebrating the Success at the End of the
Program 166
Summary 167
End of Chapter Questions.. 169
    For Discussion.. 169
    Written Assignments 170

Chapter 8 Claim Identification, Claim Management, and
Claim Avoidance 173

Late GFE 175
Defective GFE 176
Delayed Approvals or Contract Actions 179
Inappropriate Disapprovals or Comments 181
Noncontractual Direction.. 183
Flawed Technical Specifications... 185
Defective Information... 187
Claims Against You... 190
Other Considerations in Claim Management.................... 191
    Improper Use of Claims... 192
Summary 193
End of Chapter Questions.. 193
    For Discussion.. 193
    Written Assignments 193

Chapter 9 Leadership Models 195

Leadership: Getting People to Do What You Want
Them to Do...195
  The How of Leadership 195
  Rely on Influence, Not Power...197
  Examining Leadership Theory 198
  Achieving and Maintaining Respect 200
Summary 203
End of Chapter Questions 204
  For Discussion.. 204
  Written Assignments 204

Chapter 10 Communications... 205

Communications among the Team 208
What about Communication outside the Team?...............210
Communication with the Customer....................................211
A Critical Communication Skill 212
Summary 213
End of Chapter Questions 213
  For Discussion..213
  Written Assignments 214

Chapter 11 Earned Value Management............................. 215

Applying EVM Theory 215
Summary 223
End of Chapter Questions 224
  For Discussion.. 224
  Written Assignments 224
  Mini Project.. 224

Chapter 12 Negotiations... 225

Contract Negotiations.. 226
Customer Negotiations—Ongoing Contracts.................... 228
Internal Negotiations—Work Budgets 228

Support Groups—Negotiations with Support
Groups... 230
Supplier Negotiations..231
Subcontractor Negotiations 231
Summary 232
End of Chapter Questions...233
    For Discussion...233
    Written Assignments 233

Chapter 13 Coaching.. 235

Recognizing Influence in Coaching 235
Determining When to Coach 238
Summary 240
End of Chapter Questions....................................... 240
    For Discussion.. 240
    Written Assignments 240

Chapter 14 Inheriting a Program Already in Progress.................. 241

Becoming a Member and Leader of the Team 241
Importance of Continuity....................................... 243
Fresh EAC.. 244
Summary 244
End of Chapter Questions....................................... 245
    For Discussion.. 245
    Written Assignments 245

Summary.. 247
Index.. 263

Preface

Why do people go into program management? The classic answer is that program managers (PMs) derive their satisfaction from seeing a good plan come to fruition.

There is much to be said for this view, but I believe there is a more basic answer: people are attracted to program management because they have the talents that align with the discipline's needs. Just as people who lack sufficient coordination do not make a career of golf, people who lack the necessary analytical and leadership talents would naturally shy away from program management.

Individuals are fulfilled when they are able to work in areas that challenge them, but also in areas that allow them to succeed. Successful PMs love to make plans and carry them out—but much of their day-to-day satisfaction comes from solving complex problems that require both analytical and people skills that would derail those plans.

So an underlying talent for program management tasks seems a prerequisite to a happy and successful career as a PM. But beyond those basic talents, great PMs are constantly looking to learn from their mistakes and from the mistakes of others. They aspire to become better and better at leading programs to successful conclusions—identifying and avoiding performance traps; inspiring their colleagues to deeper levels of devotion and professionalism; delighting, rather than just satisfying, their customers; and, all the while, making their good plans come to fruition.

This book is intended to help early and mid-career PMs understand what they need to do to be successful. It is also intended as a resource to later-career PMs who want to get a quick infusion of another PM's successes and failures. (It is so much more pleasant to learn from someone else's mistakes rather than one's own!)

Since purely academic explanations of PM activities can be dry and snooze-inducing, the author has attempted to bring some true-to-life stories into the text. Stories are inherently more interesting than explanations. And since I believe that the PM sees the relationships and interactions of his or her team with each other, with the customer, with the requirements, and with the PM as the "meat" of the program success, the "stories" involve the people who make up the program.

The book is organized basically as a chronological "tale" of a program life cycle, with "side trips" to include important concepts of leadership, contract type, claims and claims avoidance, earned value measurement theory, and communication basics. Common pitfalls are explored in the hopes that these phrases are not heard:

- "Why wasn't I informed about that?"
- "It's time to shoot the engineer."
- "We are going to miss that date."
- "The customer is coming on Tuesday and the (blank-blank) model isn't working!"
- "How did that happen?"
- "The customer said WHAT?"
- "What do you mean Mary will not work with John?"
- "How could we be out of money already?"
- "What parts?"
- "How did I know I was supposed to do that?"

The best PMs take action to avoid the bad situations that have spawned those quotes. Even the best PMs will not avoid all problems, *but* they will find ways to mitigate the consequences of the problems that do sneak in and get the program back on track.

Just as engineers thrive on solving technical problems, PMs are fortunate to have multidimensional problems to solve: technical, contractual, interpersonal, and administrative. Every day presents a new challenge that requires a mix of drive, creativity, compassion, and insight. The best PMs are great leaders: they are people who thrive on helping others succeed, because the success of the program is found in the success of the individuals on the team.

I have been blessed to work for a great company, doing important things for demanding customers, and all the while working with talented and well-meaning colleagues. My wish for the readers of this book is to find that same fulfillment, as all leaders must, by shaping the minds and hearts of their colleagues toward excellent performance in their own personal fulfillment.

Charlie McCarthy

List of Acronyms

AC	actual cost
ACWP	actual cost of work performed
A/D	analog to digital
AW	actual work
BAC	budget at completion
BAFO	best and final offer
BCWP	budgeted cost of work performed
BCWS	budgeted cost of work scheduled
BW	budgeted work
CA	contracts administrator
CEO	chief executive officer
CFE	customer-furnished equipment
CFI	customer-furnished information
CPAF	Cost Plus Award Fee
CPFF	Cost Plus Fixed Fee
CPI	cost performance index
CPIF	Cost Plus Incentive Fee
CST	continuous self-test
CV	cost variance
DMS	diminished manufacturing sources
DoD	Department of Defense
EAC	estimate at completion
EE	electrical engineer
EMI	electromagnetic interference
EV	earned value
EVM	earned value management
FA	financial analyst
FAR	Federal Acquisition Regulation
FFP	Firm Fixed Price
FM	functional manager
FP	Fixed Price
G&A	general and administrative
GFE	government-furnished equipment
GFI	government-furnished information

I/O	input/output
IP	intellectual property
IRAD	internal research and development
LOE	level of effort
ME	mechanical engineer (or manufacturing engineer), depending on context
MR	management reserve
OPM	operations project manager
OT	overtime
PM	program manager
PO	purchase order
PTW	price to win
PV	planned value
QAE	quality assurance engineer
R&D	research and development
RFEA	request for equitable adjustment
RFI	request for information
RFP	request for proposal
RFQ	request for quotation
SE	systems engineer
SOW	statement of work
SPI	schedule performance index
SV	schedule variance
SWE	software engineer
T&M	time and material
Ts and Cs	terms and conditions
TCPI	to complete performance index
T-spec	test specification
WBS	work breakdown structure

About the Author

Charlie McCarthy, having graduated with a bachelor's degree in electrical engineering from Manhattan College, New York City, New York, joined a large electronics company. His second assignment on their Professional Development Program became permanent, and Charlie began his "first career" as a systems/digital design engineer. For most of his career, his work has been in Nuclear Instrumentation and Control systems, a field that is obviously both critical and rigorous.

As much as Charlie loved electrons and diodes, he loved working with people more, and gradually grew into technical and project leadership roles. He recognized that the source of technical and business success is derived from the work and success of the people on a program. This realization was key to formulating his philosophy of leadership as that of a "servant-leader." The best leaders put their people first, create an environment of respect and appreciation, and help them be successful in both their work and their career. Charlie aspires to be that kind of leader.

Charlie's technical and analytical interests and his interest in people find a happy overlap in program management, one of the few positions that allows an individual to work simultaneously in all three areas. Along with growth in experience and technical competence, he earned a master's of science in electrical engineering from The Johns Hopkins University, Baltimore, Maryland. He has successfully managed a wide range of projects and programs, from those with one or two people to large, complex hardware and software programs involving over 50 engineers and operations personnel. He believes in the power of motivated people working in an environment that fosters collaboration and trust and sees it as his responsibility to create that environment.

Charlie enjoys a rich and fulfilling family life. He and his loving wife, Barbara, have raised four terrific and kind children, and are currently helping raise six grandchildren (so far).

1

Overview: Program Management in the Department of Defense (DoD)/ High Technology Environment

"McCarthy!"

"Yes, boss."

"I've made some changes in my organization and you are affected."

"Oh?"

This was my introduction to program management. Apparently, I had been put into a management position and was being informed. This is not exactly how I envisioned my transition from engineer to manager to take place. I was expecting something like: "Charlie, we have been watching you, and we think you have the talent to be a leader in our organization. Would you consider taking on a management position?" Well, my vision and that of my management was not completely aligned, but this path of entry into management does illustrate one important aspect of one's career—there are a lot of outside forces that "affect" your career direction.

If you have picked up this book, it is likely that you are considering program management as a career, have been "elected" into an opportunity, or are interested in broadening your program management skills. This book is intended to address all of these possibilities, and I welcome you to the pleasant and successful use of its contents.

Program management is as much art as it is science (or engineering). Or perhaps better, it is where management skills (budgetary analysis, work monitoring, scheduling, etc.) intersect leadership traits (vision, motivation, career growth, etc.). Effective program managers (PMs) are able to combine management and leadership for the good of the program and the people that are entrusted to them. (Refer to Figure 1.1.)

FIGURE 1.1
Program management Venn diagram.

This work is challenging and often stressful, so why would one choose to follow program management as a career? Good question. There are many reasons offered: "the satisfaction when a plan comes together" is a popular one, and an important motivation for many of our colleagues. Some, sad to say, enjoy the prestige or power of the office. Some get satisfaction in successfully undertaking the complex challenges that the job entails. As a PM, it is good for you to understand your own motivation.

PMs like challenges: customers, engineers, operators (and people!), and how the nuts and bolts of getting things done intersects with the people involved. If you find this kind of multidimensional "puzzle" interesting, you are off to a good start. PMs like to solve problems: technical, perceptional, financial, and interpersonal. A good PM is a happy PM and finds satisfaction in ongoing problem solving. If this sounds like you, then you may find program management is your calling.

A PM's job is to get the work done: on time, within budget, and with high quality. This is a complex task and requires a brain-holistic approach. The left side of the brain, credited with quantitative and analytical thinking, must work with the right side of the brain, which deals with creativity and social interaction. It takes the left side of the brain to analyze progress and costs and the right side of the brain to lead a team to success.

Many leaders or managers seem to not be using their full brain, at least figuratively, and maybe actually. Right-brain managers who spend their day at their desks, reading reports, scrutinizing charges, preparing spreadsheets, or otherwise avoiding people-contact are seen as remote and uninvolved in the program. And at the other end of the spectrum, left-brain managers who do not understand their costs and schedules can lead their teams into failure and the chaos that precedes failure.

It is important to use the whole brain! The program team is relying on both the PM's management and leadership skills to ensure the team's success. It is an important job—folks on the program are investing their time and betting their careers, in many cases, on the success of the program. Large design programs can run for 5–10 years—this is up to 25% of a career span! Program management is a big responsibility.

My experience is that in large technical companies, such as many Department of Defense (DoD) contractors are apt to be, the Venn diagram tends to be shaped as shown in Figure 1.2.

This is not surprising, since in large technical organizations, PMs are often promoted to that position out of the engineering ranks. Engineers are trained and paid to be analytical, which is a great trait for a PM. But if you are more comfortable sitting alone in your office, reviewing spreadsheets (rather than talking to your coworkers), your program leadership may be lacking. I believe that you have to have a genuine love of working with people to be a great PM. Most PMs have plenty of the "science" but have little affinity for the "art."

ROLE OF THE PROGRAM MANAGER

Very often the PM has been the proposal manager (more discussion on the proposal process and its management in Chapter 5) on a winning proposal. Leading the proposal team is an excellent start for the program and is recommended whenever possible. As the proposal manager, he or she has reviewed the requirements (or even helped write them in some cases), has planned the program at least at the top level, and, importantly, has

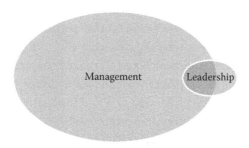

FIGURE 1.2
Management lopsided program management Venn diagram.

bought into the proposed costs and schedule. He or she is well prepared to lead the program to the vision presented to the customer in the proposal.

Executing the vision is **the primary task of the PM: getting the work done on time and on budget, with high quality, thus meeting contractual obligations and satisfying the customer**. That's the "What" of program management. The "How" is the hard part. The How comprises much of the rest of this book. Following is an outline of the "How." The PM must be all of these things:

- Scheduler/planner—Depending on the size of the program, the PM may have someone who creates, manages, and tracks the detailed program schedule. But even if there is help, the PM is ultimately responsible to see that the work is organized into a reasonable schedule and that the people on the program understand it (and ideally, embrace it).
- Progress monitor—Once the people understand and buy into the schedule, the PM must be sure that they work toward it, and that unexpected issues are identified quickly so that the appropriate recovery actions can be initiated early. The best schedules contain tangible events so that there is no question that the work is on target. Avoid, if you can, long-term tasks whose only progress measurement is **x**% complete. (Hint: Software projects are always 90% complete!)
- Coordinator—Work products that move between functional departments (system engineering, design engineering, drafting, etc.) require special attention. The product passed along must be of high quality to allow the next group in the chain to do their work effectively. And it must be on time. Late hand-offs are costly—recovery includes making sure that the delayed party is doing useful work, or is temporarily assigned to another program, so that delays (and costs) do not snowball.
- Customer interface—The PM is usually the face of his or her organization and of his or her program to the customer. While the contracts department may perform official document intake and output, the PM is the person who the customer counts on to understand the program's status and health and to make it happen. The individual customer's success is directly influenced by his or her ability to keep the program on track. Because you and the customer share the same goals, he or she can be an important ally!
- Management interface—The PM usually has formal reporting responsibilities to more senior management. It is his or her job to elevate issues that are beyond his or her control. Senior managers can

help the PM when he or she has problems, or sometimes, they can "help" by requiring unnecessary reports and meetings. Fear of this type of "help," plus occasionally a PM with a too-big ego, are how minor problems become serious problems. Careers at multiple levels can be adversely affected by delayed problem reporting. Advice: Do not let problems fester!

- Cost control—First cousin to schedule control is maintaining the program's cost within budget. It is said that if you manage the schedule, your costs will be OK—true, sometimes. You can stay on schedule by adding expensive additional hands to the team—so, yes, you may be on schedule, but you may be spending way too much money for the progress that you are making!

- Business growth—In many organizations, the PM is looked to as the source of additional business. This may be done by alertly identifying an unfilled customer need and convincing the customer that your organization is the one to do it. It might be that the customer wants new features or a more aggressive schedule (or a more delayed one). It might be that your success on a program is noticed, giving the customer the confidence to give your company more work. Every PM who has seen the heartache caused by layoffs should be committed to doing all he or she can to grow the business and thus keep his or her colleagues (and friends) gainfully employed.

- Staffing—Most large organizations use a "matrix management" structure (more on this in Chapter 7). In this kind of arrangement, the engineers, draftsmen, technicians, etc., report "solid line" to a functional manager. The functional manager's job is to support you and get you the right folks at the right time, and find new work for them when your program does not need them anymore. Effective functional managers thus provide a great service. Ineffective ones can drain the lifeblood from your program. If they give you the wrong folks ("wrong" as in being unmotivated, underqualified, slow, argumentative, uncooperative, etc.), it can waste the hours you give them to spend and can infect the esprit de corps of the rest of the team. Work with the functional manager, but insist on only the best people on the team. The functional manager must place lesser performers someplace where they will not hurt the company—but that is their problem. Your responsibility is to your program.

- Strategic planning—How can this program position the company for significant growth? Is there some way that the work can be approached so that inventions, large and small, can be capitalized upon for new work, or even to create paid work where there was none before?

- Intellectual property monitor—The PM must understand the contractual issues involved in the ownership of the aforementioned inventions. Remarkably, even if a customer pays for every hour that the engineer spends in the development of an idea, the ownership may reside with your company. Or maybe not—that is the key. Understand the terms and conditions of your purchase order or contract.

- Change manager—This is a big obligation. In complex, long duration contracts, a good relationship often exists between the customer, his or her engineers, you, and your engineers. New ideas will crop up, and your engineers, wanting to design the best product they can, and the customer, wanting the best product he or she can get, may become carried away and forget about the contract requirements. On the surface, this striving for excellence is good, but in practice it can kill the schedule and cost control. New ideas that are out of scope are good, because, properly managed, they can result in constructive changes where the customer pays for the enhancement. Even your talented engineers put you at risk—watch out for "we should be able to make this more accurate," "we can incorporate additional self-testing," and "we can provide this bell and whistle"—"at almost no cost."

- Program publicist—Your program will need support from management, from peers, and from the engineers assigned or who might be assigned to your program. Creating and "advertising" the good that your program is doing is the way to get that support. You need to attract the best folks to your program; some of the talent may have had poor experiences by being assigned to "loser" programs. The best way to attract the success-oriented engineers is to have a success-oriented program and make it be known. This may require a little public speaking and article writing for in-house newsletters. It is part of the job.

- Specification manager—Very often, on complex programs, there may be vague or interpretable requirements. Your job is to work with the customer to a win-win understanding of the contractual requirements. The better your relationship with the customer and the better you understand the technology of your program, the better you will do at this role. You will no doubt need the help of insightful systems engineers. Make sure they are your "techno-friends"—i.e., you do not have to socialize with them, but your relationship has to be collaborative, and they need to be committed to your joint success.

- Program life-cycle manager—Programs typically start small, grow to a peak, and then shrink as the work gets done or transitions into the next phase (see Figure 1.3). The PM must add and remove folks while maintaining a positive team environment, protecting the egos of those involved, and seeking career growth for folks going in and out of the work of your program. PMs must always thank people for their real contributions and need to show genuine appreciation. Genuine expression of appreciation for good work is one of the strongest motivational tools you have. But be careful—thanking folks whom the program team knows to be lesser performers undermines a leader's credibility!

- Leader—This is the most important role. The PM has the responsibility to maintain the environment so that the team feels like it is respected for the good work that it does, and so that individuals are appreciated for extra effort and outstanding results. This leadership role includes having and expressing genuine concern for the people on the program and for their career advancement and professional recognition. The PM must be the kind of person that the engineers

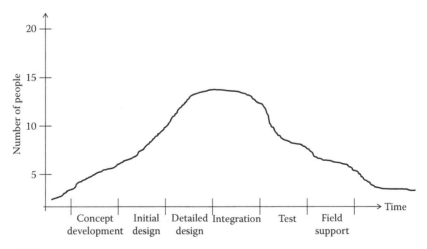

FIGURE 1.3
Program life cycle.

want to confide in and discuss ideas with. It might be stressful to have problems brought to him or her, but the PM should never imply that he or she wishes the problem had not been identified. The leader should want to know the problems and help the team solve them. And, be aware, sometimes help must be called in to facilitate or provide expert advice. The PM needs to be able to identify those times, and bring in outside experts, all the while respecting the talents and egos of those on the program.

QUALIFICATIONS, EXPERIENCE, TALENTS, AND SKILLS

How can PMs fill all these roles? Well, PMs should love the opportunity to use their skills and develop the ones in which they are deficient. They have to trust colleagues and be able to identify their strengths. They have to have enough self-confidence to realize that there are lots of people better at some things than they themselves are, and they should be able to be glad, not envious. They have to be intelligent—able to see patterns, create ideas, and collect and apply the data and information. The effective PM must be patient when necessary and impatient when the situation demands. He or she must set a work ethic example—not as a workaholic—but as a model of dedication who uses his or her work time productively and understands his or her out-of-work obligations and needs as well. He or she must be truly caring and compassionate; every great leader cares about his or her people, and the greatest leaders put them first, even at their own expense. This means that the best leaders are also willing to credit the team to the point of self-invisibility—ironically, that takes a lot of ego:

> The wicked leader is the leader who the people despise.
> The good leader is the leader who the people revere.
> The great leader is the leader of whom the people say, "We did it ourselves."
>
> *Attributed to Lao Tzu, ancient Chinese philosopher, 6th century BC*

To be an effective leader, one has to achieve the respect of his or her team. There is a residual old-school idea that intimidation breeds respect. Beware of this sociological error and the possible persistence of this old thought in your organization. Management by intimidation does not make a leader effective.

Think for a moment about what kind of people you respect and what kind of traits they exhibit. Odds are that your colleagues also admire those traits.

This is a story of which I am not proud, but it was a teaching moment for me. "Me" is the PM, and "Fred" is the lead software engineer, reporting "dotted line" to me.

Me: "Fred, I'd like your document to include words that limit our test coverage to the processor core."

Fred: "Well, Charlie, that's not what this document should focus on."

Me: "I agree it's not the focus, but this document is an opportunity to get this issue formally in front of the customer and establish a basis for our test philosophy."

Fred: "I don't agree. The document purity is more important."

Me: "Purity is nice, but we need to limit our exposure."

Fred: "I don't agree."

The conversation went on like this for several minutes, with passions rising, until Fred said:

Fred: "I don't agree, and I don't see why I have to include this in my document."

Me: "I KNOW WHY YOU HAVE TO INCLUDE IT! BECAUSE I AM THE PROGRAM MANAGER, AND I'M TELLING YOU TO DO IT!" *(This line was delivered at high volume after my jumping to my feet.)*

Fred: "ok." *(Fred exits.)*

I think: "Well, that showed him! I got what I wanted!" But after the adrenaline cleared—I think, "Wow, what a jerk I was." So I apologized to Fred—but I still got what I wanted in the document.

So here's the lesson: Intimidation might look like it works, and it is a bit of a drug. It feels good to "vent your spleen" and get subservient acquiescence to your demand. But if you think of the cost in relationships and program ownership, it is nothing but a bad approach. After my tirade, Fred might do what I wanted, but he lost ownership of the work product. He probably also lost a lot of respect for me. This is especially ironic. If one's goal is to win respect by intimidation, the real result is a loss of respect, but it masquerades as *apparent* respect—thereby utterly confusing the relationship dynamics and making them inherently phony.

FIGURE 1.4
Early program manager—"Lao Tzu."

FIGURE 1.5
Earlier program manager.

Figures 1.4 and 1.5 illustrate some early "program managers." Which one would smart people (the kind you want on your program) prefer to follow?

Qualifications

No one should enter program management without the talents that could make him or her successful ("talents" versus "skills" is discussed in the Section "Qualifications, Experience, Talents, and Skills"). Just as engineers should have an affinity for math and science, the successful PM should have the personal traits or talents that allow him or her to succeed. The personal traits that good PMs have can be categorized as "hard," "soft," and "prerequisite."

Prerequisite:

- Having integrity
- Honest
- Having a sense of justice
- Intelligent
- Trustworthy
- Industrious
- Confident
- Creative

Hard:

- Analytical
- Courageous
- Demanding
- Forceful
- Driven

Soft:

- Appreciative
- Communicative
- Collaborative
- Approachable
- Respectful
- Compassionate
- Considerate
- Altruistic
- Humble
- Persuasive

While it may seem like some of these traits are in opposition to each other, in fact they are not. As an example, one can be driven toward a goal, such as an important milestone on the program, but when a key team member needs to go home to take care of a sick relative, the PM can use his or her creativity to find another way to get the work done while respecting the personal situation of the team member. People will see the PM as committed to the goal but also as caring about them. Caring about the people in the context of the mission is the essence of leadership. People on the team see that the PM cares about others. And it is easy to extrapolate to themselves: "If I have a problem, I can expect help and consideration. If we are smart, we can be successful and still handle personal problems." It does not take too many examples of this behavior for a leader to establish the respect of the team.

Experience

Most PMs come up "through the ranks," progressing from technical tasks to being responsible for subtasks with one or two junior colleagues, to being responsible for a subsection of the design, to, maybe, technical leadership on a program. There is much to be said for walking in the shoes of the people you hope to lead. Some companies actively move people into multiple disciplines (engineering, manufacturing, test, contracts, etc.) with the expectation that this will make them good PMs. While rotational assignments may indeed be a help, it is also possible to develop into a great PM by using that "collaboration" trait that was mentioned as a "soft" trait in the Section "Qualifications." When a PM tries to understand the needs of each of the departments he or she works with, and cares about colleagues' personal success, it builds rapport that fosters direct communication. He or she will be told all one needs to know by the folks truly expert in those various disciplines. Example: If one understands that vague drawings create problems for the manufacturing manager, and cares about his or her stress and success, the PM will naturally take these concerns to the engineers who create the vague drawings, and if they care about the PM's stress, and/or that of their manufacturing colleagues, they will work to get the vagueness out of their drawings. Simple, isn't it? Thus, communication and caring can substitute for years of experience. Good news! Folks do not have to be old to be a great PM! (After all, Lao Tzu did not write down his philosophy until he was 80 years old, but we can be sure he developed it much earlier than that!)

Talents

It is important to differentiate between talents and skills. Talents are innate, and if not inborn, they are formed early in one's life experience. One can be a talented athlete, and if you are, you can develop skills in a variety of sports. But if you are inherently uncoordinated, you would be foolish to spend hours and hours practicing baseball. If you are adept at mathematics, you can pick from a variety of fields that need that talent to be successful. But if you are not—you should find a path that does not require that talent. If you do not have the aptitude for math, you can study hard and at best achieve mediocrity.

People naturally gravitate toward using their talents. Doing things at which you excel is inherently satisfying and leads to success. Good leaders recognize this and allow folks on their team the maximum latitude in finding the parts of the program where they want to work—that is where they will excel. In the old school of management thinking, one unenlightened manager might say: "John, you have poor technical writing skills, so I've decided to give you more writing tasks to build your skills." (Probably, John has low writing talent, not "skills.") So poor John does his best and, with months of practice, gets a little better at writing. But, in the process, he wastes program money and has a miserable workday. The enlightened leader says: "John, it looks like you don't care much for technical writing, but I've heard you are a great mathematical modeler." John is assigned work that aligns with his talents and is very productive and happy. Meanwhile, the technical writing is assigned to someone who wants to do it. Effective PMs understand the difference between talents and skills and manage their program with this in mind. And the best PMs talk to their team to understand the talents that they each have. No one knows better than oneself what a person likes to do. And it is 99% probable that he or she has the talent for what he or she likes to do!

Furthermore, good PMs recognize where their own talents lie and verify that they align with those needed for the job. Where they are short of one talent, they find ways to compensate—for example, relying on a colleague who has strong analytical skills. The program analyst or accountant or financial administrator can analyze program costs efficiently and find patterns that the PM might miss. The most successful PMs trust and utilize the talents of colleagues, especially in areas where they may lack those talents!

Skills

Skills are where talents are applied. As an example, you use your analytical talent to develop the skill you need to understand cost drivers on your program. As mentioned previously, it is important to one's success to have some frank self-appraisal of one's own talents. If you are strong analytically, you may do much of the program's cost analysis yourself. If not, assign it to the program analyst (or financial analyst—this function has different titles at various companies). Have him or her perform the needed analyses. Inherently, nonanalytical PMs might be able to build the skills to do this themselves pretty well, but if they spend several hours developing the skill to do it themselves, time is wasted that might better be spent interfacing with the customer or attending a technical review. So good PMs develop skills that they need and can reasonably acquire, but rely on the team members for the rest. Thus, he or she can avoid learning skills that could be so time consuming that he or she neglects work needed on other aspects of the program. The PM also needs to be alert to this second-order effect: people sometimes think that skills that they do not have are of lesser importance. This could provide a "self-justification" to pay less attention to those things. For example, a PM who is not numbers-oriented may think, "If I pay attention to the schedule, the costs will take care of themselves." This kind of thinking can lead to assigning too many people to do work to maintain the schedule, but, in the process, spending way too much money! The moral: The effective PM should pay special attention to the things he or she does not have an affinity for and get expert help in those areas.

Of course, there are some aspects of the program that cannot be delegated. Here is an important one: the ability to detect unnecessary work being done on the program and the ability to stop it without disengaging the engineers. Engineers, being who they (we) are, are always working to improve their products (usually designs). This is a good trait for the most part, but it can kill your program's finances and schedule. This is such a common problem that a program management phrase has been developed for it: "It's time to shoot the engineer!" (I always found this a little drastic, especially considering that often I was the engineer for whom it was time to shoot!) The good program manager stops "design creep." This requires two things: first, the skill to identify it, and second, the skill to talk to the engineer(s) about it. Keep in mind that the "design creepers" are probably excited and impassioned about the "improvement." Being a reasonable bunch, most engineers understand if they are told that there just is not the time or money to make such-and-such improvement. There might be some pushback, but that is where those interpersonal talents described in the Section "Qualifications" come in. I would suggest combining the following traits: appreciative, persuasive, firm, and understanding. While in the heat of the discussion, the idea of "shooting the engineer" may seem like a logical solution, you need to remember not only is it immoral and illegal, but also it is very disruptive to your program! And you may need that engineer later on anyway!

TYPES OF PROGRAMS

Programs come in all sorts of sizes and types. Some program managers tend to "specialize" or concentrate in certain areas, especially those that align with their background and experience. We will talk about several types of programs that one might encounter in a DoD or other high-technology arena.

Research

This type of program is the most "scientific" type of program. The effort might be, for example, to investigate the feasibility of the application of new technology to existing problems, and the output is likely to be formal reports or trade studies. Because there are few disciplines involved (no manufacturing, supply chain, etc.), this type of program has fewer hand-offs and internal interface challenges. Here, the primary requirement is to produce results that are satisfactory to the customer. The best way to achieve this success is to make the customer a collaborator as well as a customer. The more intellectual content he or she has in the program, the more likely he or she is to be satisfied with the results. Collaboration at all levels, between the customer's organization and yours, is generally positive and to be encouraged. There are risks, however, in this collaborative relationship. It would be easy for the team to stray off topic and produce results that are not needed or wanted by the customer's upper management. In this case, everyone loses. The effective PM on this type of program should take every opportunity to check the direction and progress with the customer's management at all levels.

A research-oriented type of program also brings the risk of cost over-runs. Because the results of the work are hard to quantify or may not even be achievable, very often this will be a "cost-plus" or "time and material" (T&M) program. These types of programs are characterized by the customer paying for all the costs incurred on the program. But this type of program is by no means a "blank check." In fact, the customer is likely to provide more scrutiny toward the money being well spent since there are less-definitive objectives. This type of contract will often be written with deliverables such as reports or presentations. Consider, for example, a simple program with a single report at the end. The contract may be more or less prescriptive, with clear definitions of what should be included, or maybe not! The strategy should always be this: Please the customer, but with an attempt to delight the customer. In this case, the fixed variable is the cost (contract price, less profit and overhead), and the variable is the quality of the report. It is usually easy, if the PM is attentive to the costs, to meet the cost target—one simply scales back on the depth of the report if the funds are running low. The challenge is to make sure that the content is of high quality. Since it is difficult to measure how many hours might be required in the research that leads to the report, the best way to ensure satisfaction is close coordination with the customer. Weekly or monthly phone calls (depending on the size and duration of the program) are often the best way to achieve this coordination. Sometimes the contract may call for these reviews, but it is suggested that, even if it does not, this is a great practice to ensure the customer knows how the work is progressing and where his or her money is being spent.

Design

A design program can be of any size as well. Often, the output of this contract is a collection of drawings or software products that meet a customer's specification. Sometimes the contract requires the fabrication and testing of a qualification unit, to verify that the design meets the requirements of the specification. This is one of the more challenging types of programs. One cannot "adjust" the content of program as discussed above as funds get depleted—the contract requires that a unit be built and tested and that the equipment passes the tests! Because of the uncertainty of the work (and as a function of the technical challenge of the design), these contracts are often some variety of "cost-plus" contracts. This reduces your company's risk, of course. Many contracts are written with cost incentives, or with absolute caps, or with cost sharing provisions for overruns. This type of contract mitigates the risk but does not eliminate it. Company senior management is almost certain to expect full profit (no overruns). The customer will also receive credits or demerits from his or her management based on how successful the PM is at delivering a successful design on time and within cost.

Here it is vital to have sufficient technical understanding of the program to determine if it is on track. The PM may need, but certainly should have, technical advisors that help in this determination. They must be chosen to be insightful, realistic, and painfully honest and direct. When (maybe "if" but most probably "when") things are going poorly, the PM has to alert senior management. Most likely they will provide "help." Depending on how talented they are, it may be real help, or it may lead to additional reporting requirements, which may actually be a distraction to understanding and fixing the problem. Nonetheless, the PM must never fail to identify budding problems—nothing is worse than multiple levels of "surprises" that are not understood to be challenges that are being addressed. (I have seen five layers of management replaced, shuttled off, early-retired, or subjected to some other unfavorable career event because problems were not brought forth in a timely way.) This is where that "courage" talent would be valuable.

Production

Once the design is done (by your company or by others), the next logical step in most cases is production of the product. Production programs have their own challenges, but, in general, contain fewer risks than design programs. Because of the reduced uncertainty, many of these contracts are likely to be fixed price. Because the material content is known, and because there has probably been some experience in measuring the labor content during the preproduction phase, costs are more accurately determinable. Fixed-price contracts carry their own risks, principal among which is the opportunity to lose money—maybe a lot of it. Since it's the PM's job to make money for the company, not lose it, these contracts require their own type of care. Primary issues here are personnel and material availability. You cannot build equipment on schedule if you do not have the parts, and you cannot put the parts together if you do not have the proper workforce. A shortage of either can cause schedule delays, which will either aggravate or infuriate the customer, and will lead to program costs as the support team (quality assurance engineers, contract representatives, even the PM!) will be on the program longer and charging to it. Schedule control is thus vital to financial success, and therefore it must be watched closely. Your manufacturing team can advise you on how (and if) to work around certain part shortages to keep everyone busy, but you must also be alert that some of your team may have some potentially conflicting goals—if there is no other work in the factory, the functional managers might prefer their folks to be underemployed on your program than to be idled and charging overhead.

Because the various types of programs carry different risks, they are properly controlled by different types of contracts. Thus, frequently, but not always, a contract type consistent with the nature of the program is employed. These types are outlined in the next section.

TYPES OF CONTRACTS

Customers have a wide variety of vehicles with which to contract for work with suppliers. The different types of programs described above align with certain types of contracts, wherein risk and reward are managed appropriately for the work contracted. Here are the major types:

- A "cost-plus" contract is one in which the supplier's "cost" is paid, "plus" a percentage profit fee. The labor costs and material mark-ups are subject to disclosed rates, audited by the government to be fairly collected, with strictly allowed and disallowed overhead costs distributed in an auditable way. Here, all the costs accumulated are reimbursed by the customer, along with a pre-negotiated, or an incentivizing fee. There are three major types of cost plus contracts:
  - Cost Plus Fixed Fee (CPFF)—In this type of contract, costs are reimbursed at an agreed to rate at the time of order placement. For example, if a prototype radar is to be developed, the customer may pay what it takes to do the work, and allow a prearranged fee for the service.
  - Cost Plus Incentive Fee (CPIF)—In this type of contract, costs are reimbursed, but better performance (early delivery or reduced cost, for example) will invoke a higher percentage fee. For example, if the work gets done earlier, this may be an advantage to the customer. And to "incentivize" the supplier to put the extra resources on the

program that would lead to earlier completion, the contract value is adjusted upward to make it worth the supplier's extra effort.

- Cost Plus Award Fee (CPAF)—In this type of contract, the fixed fee is augmented by increased performance on some key parameter(s) of the effort, for example, an airframe's speed or fuel economy, or a computer system's throughput or latency.

Another variant of the CPFF contract is one wherein the fee is fixed at a certain percentage of the nominal value of expected costs, thus attempting to control cost overruns. For example, if a program is estimated to cost $10M, the fee might be set at 10% of the $10M, or $1M, for a total price of $11M. If there are cost overruns, and the fee is fixed at that $1M, then the contractor's profit percentage erodes. If the costs were to rise to $15M, for example, the price grows only to $16M, and the fee as a percentage shrinks to 6.67%. Because a company's performance is often measured in profit percentages, so too is the program's performance. You may have made just as much profit in dollars as if the costs were controlled at $10M, but you used company resources that could have been making 10% profit to make your mere 6.67% profit. Thus, you (and your company) have a very good reason to control costs toward the nominal value.

While cost-plus contracts in all their variants are low risk (in that the company cannot lose money on them), they still must be managed well. If an anticipated fee is reduced because of inferior performance, there is a significant "disappointment" in the senior management team. If the profit shrinks to 3% or 4%, your program, especially if a large program, can dilute the profit percentage of your group or of the division significantly. And thus, even if the company does not actually lose money, that poor performance is a PM nightmare.

- Fixed price—This type of contract is just what it sounds like: the price is fixed, (no matter what it costs to execute). Sometimes they might be called "FFP" or Firm Fixed Price, which serves to emphasize the fixed-ness. It should be used for work that is well defined, such as production runs of proven designs. These are low risk because the costs are easily identified and estimated. But because this type of contract reduces the buyer's risk of paying more money for a product, it is often favored by buyers where the risks are not really that low. For example, a system may successfully pass preproduction tests, but may prove more difficult to manufacture in quantity than expected—which is due, for example, to component variation or DMS (diminished manufacturing sources—aka obsolescence). Because the risks are higher to the contractor, this type of contract typically carries fees that are significantly higher than T&M contracts—possibly in the 15%–20% range.

- A subtle variation on the FFP is the Fixed Price (FP), also called the "Lump Sum" contract. It is product oriented, and as the name suggests, results in a single contract price for the product.
  - A variant of the FFP contract is the "Fixed Price Incentive" contract, which is basically FP, but has provisions if the costs are higher than (or lower than) expected. For example, a program is supposed to cost $10M and is priced, say, at $12M (20% profit). But something goes wrong, and the actual cost is $11M. Here, the overrun ($1M) is "shared"— perhaps the customer has agreed to a 50% overrun share. So instead of making a profit of only $1M, in this cost-shared contract example, the profit would be $1.5M. The idea is to incentivize the contractor to control costs, but not put him or her in a poor financial condition for issues that may not be knowable at the time of contract placement. Often, underruns are shared too, so that if a program goes better than planned, the buyer would get some of the excess profit back by way of a reduced contract price. Figures 1.6 and 1.7 illustrate cost overrun and underrun scenarios and

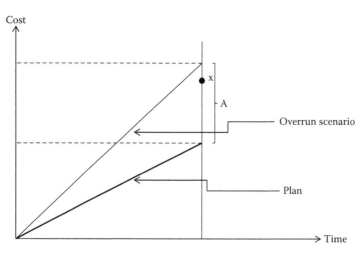

FIGURE 1.6
Overrun scenario: Cost overruns A will be shared up to point X, the "point of total assumption."

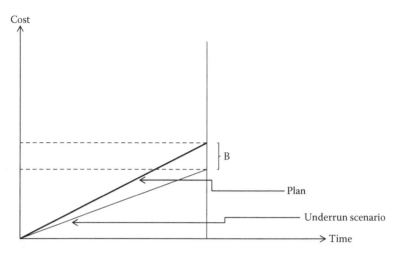

FIGURE 1.7
Underrun scenario: Underrun costs B will be shared back—oftentimes there is no downside cap.

the cost-sharing concept. Often times, these incentives have limits, so that the ultimate cost to the buyer is capped. Woe to the PM who manages a program that exceeds the protection!

• T&M—This type of contract is best suited to programs where it is not possible to accurately estimate the amount of work needed to complete a work scope. Research work, in which it may not be clear that a particular problem even can be solved, would be an excellent candidate for a T&M contract. Here, the customer is actually buying work ("time"), usually in terms of engineering hours and prototypes, demonstration models, etc. ("material"). Since the outcome is difficult to predict and estimate, this is a particularly equitable approach for "highly developmental" or particularly challenging work. Objectives may be set, but contractually they are treated as "goals." Payment is based on hours spent and material purchased, all at a pre-negotiated mark-up for labor costs and material costs. Frequently, bills are issued monthly. This is a particularly low-risk contract to have, since all the work done and materials purchased will be paid for in this type of contract. In recognition of this low risk, the profit or fee on this type of contract is usually low—maybe in the 8%–10% range. The biggest risk is that the customer will become dissatisfied with the outcome or progress and will terminate the contract.

As you can see, there are two basic types of contracts—cost-reimbursable contracts (T&M, CPFF, CPIF, CPAF) and fixed-price contracts. But it is obvious that many different variants, or hybrids, can be constructed from these two main types, selecting features that are peculiar to the individual types.

Of course, the contract type and, for incentivized or T&M contracts, the fee are subject to negotiation. But in competitive bidding, often the type of contract is established by the customer, and all bidders are forced to bid under those terms. The bidder can recognize the risks in the contract and bid a higher fee, but that will make the price higher and thus may lead the bidder to lose to a more aggressive (or more naïve) competitor. Most large companies tend to be risk averse—winning a big contract that overruns the bid price can produce years of profit drain. Thus, on fixed-price, or low-level or low-capped incentivized contracts, a proposal manager may expect a great deal of scrutiny in verifying that the costs were accurately estimated.

At times, companies may "buy in" to design contracts in the hope of winning exclusive or at least favorable position on the resulting production contracts. This can be very risky, and possibly illegal under the FAR (Federal Acquisition Regulations). Costs must be stated accurately, even in competitive bidding. However, quotations can be optimistic or pessimistic, and fees can be lowered to 0% if the business advantage is large enough. But suppose that the company does aggressively bid and win a design contract. The follow-on production contract would likely be competitive, and the designing company may not have that large an advantage (from its initial build). In fact, maybe it does know how challenging the production is, and a competitor could win at a lower price because he or she is not aware of the pitfalls. Or, the program can be canceled by Congress with minimal or even no production. The gamble of "buying in" (even within the limits of the FAR) is so high that most large companies simply will not do it (knowingly, anyway).

The bottom line is that it is important that the program be properly funded under the correct type of contract. This is one reason that the proposal manager is often selected to be the PM. The proposal manager is responsible for identifying the risks in a bid and taking mitigating actions to make the risks reasonable and consistent with the profit level. He or she should have a vision of how the work can be performed within the dollars available and should identify those risks and mitigations to senior management before the decision is made as to whether or not to proceed to prepare a proposal. Thus the PM "owns" the program from the very start—and knowing that as proposal manager he or she is likely to be the PM—he or she is unlikely to take unreasonable risks just to win the contract award!

ORGANIZATIONAL OVERVIEW—DEPARTMENTAL INTERFACES

One good analogy for the PM is to see that as the conductor of a symphony orchestra. All the sections (strings, percussion, woodwinds, etc.) must work together to make beautiful music. So if the PM does not have a violin section, but he or she does have engineering, contracts, finance, logistics, drafting, materials management, operations, and quality assurance. His or her job is to have these diverse sections make the beautiful music that comes from a successful program. In most large organizations, the roles of the departments are pretty well defined by procedures and history. But on any particular program, the talents of the people assigned from these organizations can vary. Smart PMs select team members that are smart, skilled, talented, cooperative, experienced, and collegial. But, you know, not everyone has all of these traits. In any good team, members of the team who are strong in one area should help out those who are weaker. The PM is both the leader of the team and a member of it, and he or she needs help too in areas where he or she may be weak.

Suppose you are a PM on a design job and despite your best efforts, you have been assigned a contracts administrator (CA) who is weak in understanding intellectual property (IP) issues. Fortunately, you are experienced in IP matters, so you can help the CA watch out for those issues. And you do it constructively and helpfully so that the CA knows to bring these issues to your attention; if you demean the CA even once for not knowing what maybe he or she should, you can be sure the CA will not come to you again for help.

One of the most important roles that a PM can perform in the program is ensuring that all the disciplines (but actually the people in the disciplines) are working together constructively. When assembling your team, it is best if you can find people who have worked together harmoniously before or have mutual respect through casual or secondhand association. In many organizations, the PM is not able to make the selection absolutely, but he or she can help decide in conjunction with the functional manager of the discipline. Nonetheless, do not accept anyone who is not ethical, responsible, or cooperative. One "clunker" on your team can demoralize the other members of the team and lead to reduced ownership of the program. And, it is definitely worth the extra time to check in with team members about how they would feel about working with a candidate you want to add to the team. Watch out for (and probe deeper) if you hear "well, I guess he or she might be OK...." The dynamics are absolutely amazing if people cannot work together.

I (the author) had a program one time, and I noticed that person A had information that person B needed to do his job, and person B complained to person C who got the information from person A. I thought, "That's odd," and I asked person C "What gives?" He said, "Don't you know that Person A and Person B hate each other and haven't talked to each other for years?" Well, no I didn't. Worse, I don't know how long that was going on before I happened to notice. I was absolutely astounded—I knew both A and B and thought well of both of them. BUT that was not enough! Now when I form a team, I say to the candidates, "We are thinking of X, Y, Z, W, and P for the team. How does that sound?" This has the double benefit of avoiding putting people who don't get along on your program as well as showing respect for everyone's input and showing that you care about their opinion and welcome their advice. I learned the hard way—hopefully you won't have to!

In theory, the interfaces between the departments (and people) on your program are well defined by procedures. But people are people and the best PMs make sure the whole environment, including sticky parts of it, like personalities, is conducive to getting the work done efficiently.

SUMMARY

In this introductory chapter, we considered the essence of program management, which is easily divisible into its "management" and "leadership" constituents. PMs can usually expect a great deal of guidance in the management portion from established policies and procedures at their organization, and therefore may need to pay extra attention to the leadership portion of the function. Effective program management requires that the PM takes on various functions effectively—roles, such as customer interface, progress monitor, performance coach, planning and scheduling, etc. The qualifications, talents, and skills needed to be successful are explored.

Programs come in various types, and are managed through different types of contracts. The various contracting vehicles (e.g., cost plus, fixed fee) and their variants are summarized. Ideally, the customer and the supplier agree on the correct type of contract to govern the particular type of program, and it is part of the PM's responsibility at bid time to insure that the contract type is appropriate for the nature of the program. To successfully execute the program in the PM's organization, we discussed the importance of and the ability to select and harmoniously engage specialists from the various functional organizations of the company.

The remainder of this book provides more detail on the PM's role and the environment in which he or she must work. The nuts and bolts of the work are peppered with real-life examples that illustrate the science of program management and the art that is necessary for success.

END OF CHAPTER QUESTIONS

For Discussion

1. What aspects of program management do you find:
   a. Challenging?
   b. Exciting?
   c. Boring?
   d. Distasteful?
2. How important is experience in being an effective PM? How important is technical knowledge of the program?
3. You have been assigned to manage a "cost-plus" contract. Do you think the fact that it is "cost plus" will affect how the functional managers would like to staff your program?

Written Assignments

1. Think of two leaders from your personal experience, one "effective" and one "ineffective." Contrast their traits in terms of those discussed in this chapter and evaluate which traits contributed to their success or failure.

2. What kind of program would you like to manage (e.g., design, production, research)? Why does this type appeal to you?

2

Learning the Ropes: Understanding the Culture, the Customer, and the Program Capabilities

In this chapter, we will discuss the environment of the program and how an effective program manager (PM) can use it to his or her advantage. We will also cover how the program environment can be used to avoid problems that can derail the project.

Think for the moment of the program as a system that is living within a set of other systems—the Company, the Customer, and the Team. Figure 2.1 illustrates that intersection of systems.

At times the program will interact individually with each of these regimes, but it always is a part of these three. The better the PM understands the expectations and cultures of these three overlapping forces, the better he or she will do at using those forces to foster the success of the program. Let us consider each of them in turn.

THE PROGRAM IN THE COMPANY CULTURE

The PM was chosen by the Company, thus indicating a level of trust in the individual. Despite management oversight, only the PM is able to accurately and promptly determine the health of the program and is in the best position to diagnose its ills and take action to cure them when (not if) they arise.

The organization will provide a set of procedures and processes that are intended to help the PM and provide visibility to senior management. Some companies are in love with metrics, but oftentimes it is a passing

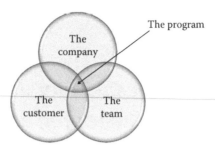

FIGURE 2.1
The program in its natural habitat.

flirtation. Metrics can be very helpful, but can also be deceptive. One of the basic concepts is earned value (EV). EV is the estimate of how much progress has been made as money is spent.

A simple example: Let us suppose the only work a project has to do is to deliver three documents, and let us suppose all three are of equal effort and will be done sequentially. If the project has $100K to spend, for example, $33.3K will be planned for each item. Work begins, and at the completion of the first document, the costs incurred are only $30K. So while only $30K has been spent, the EV is $33.3K. The second document also goes well, and it too is done for $30K. Good: Although only $60K has been spent, the program has *earned* $66.6K. But, surprise, the last document is problematic, and it takes $45K to earn $33.3K of EV. Overall, this is an overrun project, and $105K ($30K + $30K + $45K = $105K) has been spent to earn only $100K of value. Often, the amount earned is divided by the amount spent, and a cost performance index (CPI) is computed. Thus, a CPI of unity is right on target for accomplishing versus spending, while CPIs larger than 1 are good, and CPIs less than 1 indicate that you are spending more than you are accomplishing (or "earning").

Of course, in the real world, the calculation and maintenance of EV is very complex—instead of three sequential tasks, there may be 50, many of which may overlap in time. And if an EV is computed monthly, work in process at the time of the calculation must be estimated. Thus, the calculation of an accurate EV is difficult and prone to abuse. Suppose an engineering manager has a $100K task in progress that is not doing well. He or she may be reluctant to say that he or she spent $50K, but thinks he or she might only have earned $40K. If it is hard to tell, he or she might be inclined to optimistically estimate $50K of performance, hoping to make it up, or hoping that he or she has been too pessimistic in his or her assessment of progress. Maybe he or she has not heard that "hope is not a strategy."

If your project is one that has long-term tasks and must rely on estimates of EV from the folks doing the work, you are at risk. This risk can be minimized by planning shorter-term tasks, but also, even more importantly, by establishing a team culture in which the engineering manager is comfortable telling you what he or she really thinks. Cultures that "punish" or "hold people accountable" are prone to the problem of overly optimistic estimating of progress. Management (you included) cannot solve problems if they are hidden.

Cultures that are punitive cause people to varnish the truth. Often this varnished truth is mistaken for lying—but usually, it is more that the manager hopes that things will get better before he or she has to report negative news. That is why successful PMs seek and get open and honest communication. Your subordinates will tell you the bad news *only if they think it will help to tell it to you.* They will use the "hope-not-a-strategy" as long as possible if they think you will bite their heads off for reporting problems.

The scene: A senior level staff meeting, where PM A is attending. At the end of the staff meeting, the Big Boss goes around the table, and each of the PMs reports any news of which he needs to know. Unfortunately, the PM sitting to the left of PM A reports that she is behind schedule, and there is a chance of missing an upcoming delivery date. The Big Boss is concerned, so "concerned" that he becomes visibly (and audibly) angry. In fact, the diatribe is so heated that when he concludes he is red-faced and huffing and puffing. Then he asks, "PM A, do you have anything to report?" PM A responds: "No, sir—I saw what happened to the last guy." Everyone laughs, including the Big Boss. And the tension in the room is immediately dispersed.

There are at least two lessons from this vignette:

1. Humor can be used to diffuse a tense situation.
2. The real lesson: Negative responses to bad news have the absolute effect of reducing the reporting of bad news.

If the Big Boss had said—"Gee, that's scary. What can *we* do about it?" he or she would have expressed his or her concern, and gotten more information—precisely what he or she would need to be able to help! Always welcome both good news and bad news—you need to hear the bad news if you are going to help!

So, metrics can be helpful, but because of errors in establishing values for tasks and errors (or optimism) in estimating their completion, they are only one tool in your program management toolbox. In fact, it may be the most overvalued and misused one in there. The best "metric" is talking to the people responsible for the work. An honest "How is it going? How can I help?" is the very best tool a PM can have.

A critical, multimillion dollar project is launched at $t = 0$. Getting the right folks assigned quickly puts the program behind at the start. To make up time, more folks are added, and the spending picks up nicely. But because of the inefficiencies of people not familiar with the program and each other, the accomplishment does not match the money spent. However, it is early in the project and no one wants to think that they cannot recover—so estimates of EV are optimistic. That would not be too bad if the problem was recognized early and action taken. But, even though the program leadership and the engineers on the team are working hard and doing their best, the program continues to slip. But the "stoplight" indication of EV was still "Green"—it remains green for several months. Then, the first tangible outputs—reports to be submitted to the customer—are due, and they are nowhere near ready. Green status skips Yellow and goes to Red, immediately. Surprise! The PM and multiple levels of management all have their careers adversely impacted as new leaders are installed. The capital crime of "Surprise" has been committed. And why? Was anyone sleeping on the job, lying about what they were doing? No. The surprise came because the reporting was based on EV estimates that were overly optimistic—either because of fear of reporting bad news, or because the managers were estimating optimistically and hoping (not strategizing) that things would get better. The way to prevent these career-bending scenarios? Talk to the people and have them trust you well enough to give you the bad news as well as the good. Also, be forthright in telling senior management that the program is having trouble (and what you are doing to resolve the issues!)

The Company also provides detailed procedures that tell all the functional organizations on the program exactly what their responsibilities are. In general, procedures are subject to interpretation, and very often are not read until something goes wrong. People tend to do what they have always done. This means that as PM you should not rely on people knowing and performing their responsibilities, just because it is in the procedures. Again, similar to assessment of progress, the way to stay out of trouble is by talking to the people on the project. You should expect that things are being done as they should be done, but, <u>without accusation</u>, you need to <u>verify</u>

that they are being done. Why "without accusation"? Simple—if you talk to people, and they feel they have to defend themselves, they will give you only part of the truth. It's unlikely that they will lie, but it is very likely that they will prevaricate if they think you are going to react badly. As a leader, your approach should always be: "How is it going? How can I help?" not "What did you do wrong now?" If you really want to help, not pin blame, the team will learn that you are not punitive. They will trust you. You will have the information you need and thus not be "surprised." And then, very importantly, your program will not surprise senior management.

THE PROGRAM AND THE CUSTOMER (AND HIS OR HER CULTURE)

The PM is the primary interface with the customer. It is important to understand the customer's culture and work responsibilities, and report to him or her accordingly. Most often, the PM's prime contact at the customer's office is a PM also. The customer's job is to make sure that the program goes as planned: on-time delivery of a high-quality product within the costs allocated. Your customer's success is bound up, therefore, in your success—your challenge is to make the relationship collaborative. Customers who are more of a taskmaster than a collaborator can make it difficult to set up a mutually supportive arrangement, but even if the relationship is antagonistic, you do, in fact, share the goals of cost control, schedule control, and quality of product. Let us take these three shared goals and examine them one at a time.

Cost

Of course, on a fixed-price contract, cost is all your problem—basically. "Basically" because even on fixed-price contracts, cost overruns can affect your customer. If your program gets in trouble, it is very likely that you and your management would look for reasons that the customer might share the responsibility. Maybe you received incomplete or faulty information; maybe you did not receive government- or customer-furnished material on time or it was defective. Maybe software that the customer provided did not work as planned or was "flakey." So, good customers want you to succeed, not only out of altruism, but also because it reduces their risk of these "maybes" as well.

On time and material (T&M) contracts, cost control is a shared responsibility with the customer. While the financial risk to your program is reduced, remember that that risk is actually transferred to your customer. He or she has to make sure that he or she is getting what he or she is paying for; therefore, remember on these contracts, which are T&M precisely because it is difficult to quantify the deliverables and the amount of work necessary to produce them, it is inherently problematic to gauge progress. It is important to make the customer aware of your progress, and set up ways that he or she can be successful in monitoring your progress. For example, if the deliverables are sparse and represent only a portion of the actual work, it might be a best practice to provide a monthly progress report so that your customer knows what your team is doing and can easily report your progress to his or her management. Here, too, the cardinal sin of the surprise applies, but rather than surprising your management, you run the risk of surprising the customer (and your customer's management).

The key preventative medicine against the surprise is good communication. That progress report might be the best way to let the customer know of problems (and what you are doing to resolve them). Not that verbal communication is bad—actually in most cases, it is much better to discuss problems as they arise. Your progress report is for "the record" and also supports your customer in his or her role of demonstrating that he or she is managing you appropriately. Writing down the bad, as well as the good, is healthy for your relationship with the customer, and thus for your program as well. Although, on a T&M contract, running out of money before the work is done is a contractual possibility, and asking for more money (with justification) is theoretically acceptable, to not provide warning that you might need more to finish the job is unforgivable. Remember, your customer's management is just as anti-surprise as yours is. If you surprise your customer, and in turn he or she surprises his or her management, everybody loses.

Very often, the PM will be dealing with long-term customers of the organization. This is a large advantage, because understanding the customer and his or her hot buttons is one of the tools the PM has to help avoid those surprises! At times, customers can be very helpful in resolving even the stickiest cost problems. He or she may have the latitude to change the contract type, extend the delivery schedule, or otherwise (with your justification of why it is equitable to do so) make other arrangements to get your program out of hot water. Remember, though, that the customer is obligated to do his or her job well and with integrity, and just because he or she might *want* to help you is not enough. Your customer does not and cannot pay you for poor performance. But often when things go badly, there is "blame" on both sides, and, therefore, there is a joint motivation to resolve issues to both sides' satisfaction. Your goal is to leverage the trust and rapport you have with your customer in order to find creative solutions to the program's problems—*as early as possible!*

Suppose your Department of Defense customer is contracted to deliver a piece of government-furnished hardware by a certain date, and fails to do so. Contractually, you are entitled to cost reimbursement and schedule relief. The smart PM, however, does not start calculating how much can be made from his or her customer's misfortune. The smart PM figures ways to work around the delayed item with minimal impact to the program. If sufficiently creative, the PM may be able to minimize the impact to something quite negligible. This will prevent the customer from suffering the embarrassment of causing a "claim." Customers are people too, and if you help them every time that you can, it is very likely that they will be amenable to helping you. Think of your customer's problems as opportunities to help him or her, rather than an as an opportunity to "stick it to 'em."

Schedule

Because your work or product is most likely part of a bigger task, the orderly progress of your program, including and especially schedule performance, is vital to your customer's success. Therefore, just as in cost control, discussed in the previous section, you and your customer are partners in the schedule success of your program. Maintaining schedule commitments while controlling costs and maintaining quality is no easy task. There are dozens of variables and trade-offs to deal with events that might derail schedule performance, but strategies to resolve many of these will require the cooperation of your customer. The "shared goal" concept is a powerful way to encourage customer support for strategies that keep you on schedule.

Quality

The wrong way to manage schedule and cost is to compromise on the quality of your work or product. Sometimes it is possible to meet the letter of the specification but not meet the real need of the customer. Specifications, like everything else, are not perfect. While every case is different, it is important to meet your customer's needs even if you could legally "get out" of doing so by meeting the letter of the contract. While at times, if your relationship with your customer becomes adversarial, you may need to remind him or her that you can fall back to strict conformance (sometimes called "vicious compliance") if you need to, it is far better to never get into this situation. Remember that a schedule problem is bad, but a poorly performing system is far worse. But most of the time you and your customer, recognizing your partnership in quality as well as cost and schedule, will work together to maintain quality despite temptations to take an easy way out. When you compare cost, schedule, and quality imperatives, quality is king—quality, far more than cost or schedule performance, has strategic impact on you, your program, and your company.

In summary, PMs need to be problem solvers and need to use all their tools, including customer rapport and mutual creativity, to solve problems just as fast as they arise.

THE PROGRAM AND THE TEAM

The PM is the team leader. The team's internal culture is one of the best predictors/precursors of success. To be successful, the team has to work together with respect and trust. They cannot be second-guessing each other, since that breeds hard feelings, defensiveness, and retaliation. That is why it is important to know who "just can't work with" whom, as we discussed at the end of Chapter 1.

And remember, the PM is a part of the team as well as its leader and has to be sure that his or her relationships with the team are excellent. This is achieved by well-placed humility and a servant-leadership concept. If, as PM, you think you are a better electrical engineer (EE) than the engineer on your team, you are in a bad position. Maybe you need a more qualified EE, or maybe your ego is so large you cannot see around it. If you think you are such a great EE, maybe it is time to change roles! Respecting and appreciating the work of your team is the first step in earning their confidence and trust.

Of course, you will have to identify and deal with underperformers. The best way to avoid this problem is to insist on talented, motivated people, already checked out by their peers, when forming the team. It is way easier to wait a few days for the right person, rather than settle for someone who will underperform and demoralize the rest of the team as well. And as painful as it may be, get any underperformers reassigned to other projects or roles where they can be seen as contributors. It does no one any good to "carry" marginal folks—it is very likely there are assignments where they can thrive and be appreciated.

Be alert to unfair criticism among the team. Your job is to get the work done in the most efficient way possible. Peer critique is key to getting the best product—but watch out for the "piling on" syndrome. If a review becomes too critical, you must step in and deal with it. Allowed to continue, the disaffection of a single team member hurts all the team and puts everyone on edge. Never tolerate what appears to be disrespect or meanness—even if you are wrong, and the interaction is not mean-spirited, it is worth stopping the action and checking. People are different: some are direct to the point of rudeness, and some people are sensitive to the point of oversensitivity. No matter—do not let it adversely affect progress. Talk to the person, depending on the situation, in private. Set a standard of respect and appreciation that is exemplary, and your team will respond in kind.

But most of the time, if you have successfully selected your team, you may not have to do much at all in terms of managing their production or behavior. Good PMs know when to keep their hands off and when to become involved.

Across the spectrum of too-much versus too-little management, the more common problem is the former. Micromanagement saps the team's ownership of their work. If they become your minions, you might feel powerful, but they might feel like they do not have to think or work as hard, and together you will go smugly into failure. Giving people the freedom to do a job their way always—well, almost always—heightens the chance for success. Yes, there are exceptions, and you have to know when to insist on one particular path. But remember that even if your way is inherently 10% better, if you force it down the engineer's throat, his or her heart will not be in it and that 10% better will be lost in short order—and if things go wrong, you can be sure that you will be blamed. The impetus for your engineer to see that your idea works is much less motivational than seeing his or her own idea work; and thus, the drive to have it succeed is reduced. And the extra effort that your program might have gotten will simply vanish.

Once a young PM led a 30-person project, complete with systems engineers, EEs, mechanical engineers, and software engineers (SWEs). In addition, there were support organizations: contracts, finance, senior management, and later, operations. At that time, the development of software was a chronic problem. Software completion reached 90% quickly, but, of course, that last 10% took forever. The PM did not know too much about software development and was a little scared. But then, a miracle happened. For logistical reasons, the software team (eight people) was sequestered in a remote part of the building. The PM would check in on them periodically, but he left his whip holstered securely. He had a sense that everything was going OK—and it was. The software folks, left to their own devices, formed a cohesive, high-functioning team. Senior management was not too comfortable with this "hands-off" approach; in fact, the standing joke was that we should get a full-sized cardboard cutout of the PM and place it in the lab. But, despite (or perhaps because of) this allowance to form the team, the SWEs made every delivery of every work package on time and with excellent quality. When the PM peeked under the hood, he saw that when a work package was due, other members of the team pitched in and helped, even at the expense of their own commitments. (They reasoned that if they needed help later, they would get it—and they did.) Knowing when to trust the team is one of the most important skills a PM can have—second only to identifying that there is a problem brewing and the team cannot handle it alone—ideally, they will tell him or her!

Thus, as PM, you must understand the forces coming from the three regimes of Company, Customer, and Team, and how to maximize the pluses and minimize the minuses. Some examples of these dynamics follow in Tables 2.1 through 2.4.

TABLE 2.1

The Team Interacts with the Company

Company and Team	
+	−
Company-wide appreciation picnic	Increased deduction for health care
Improved rules for tuition reimbursement	Negative changes in overtime policy
New senior manager, great leader	New senior manager, poor leader
Promotion of a team member (congratulations)	Promotion of a team member (jealousy)
Company supports team member in paper presentation at conference	Company does not have funding to allow attendance at appropriate professional conference

TABLE 2.2

The Company–Customer Interface

Company and Customer	
+	−
Customer places additional work with the company	Another program for that same customer gets into trouble
Company announces lower rates for next year	Company announces higher rates for next year
Customer senior manager visits company	Customer calls senior manager to a meeting to discuss the company's performance
Admiral/General/CEO visits plant	Admiral's visit requires lab "clean up"
Admiral stops in the lab and chats with team	Team is ready with demo, but Admiral does not arrive

TABLE 2.3

The Team Interfaces with the Customer

Customer and Team	
+	−
Customer writes letter of commendation to team	Customer writes letter of concern regarding team's performance
Customer visits and has productive meetings with team	Customer visits and has counterproductive meetings with team
Customer invites team to tour test site	Customer disapproves document for invalid reason
Customer places phased order promptly to ensure continuity of the team	Customer delays placement of continuation order, and team members are reassigned
Customer facilitates Internet access for visiting team members	Team has limited or no access to Internet when visiting the customer's office

TABLE 2.4

How the Customer, Company, and Team Interact

Customer, Company, and Team	
+	−
Customer sends letter of appreciation; team gets bonus	Customer sends letter of concern; team gets reprimanded by senior manager
Customer comes and gets picture of himself or herself with team	Customer heavily edits or will not permit article on the program in the in-house newsletter
Customer places order early to allow for extended legal review	Order placement for next phase gets lost in legal review
Because of the team's success, customer places more work with the company	Customer cites team's poor performance as a reason not to do more business with the company
Customer and the team file joint patent application	Intellectual property issues prevent cooperative patent filing

While some of these forces are clearly out of control of the PM, the reflection of these events can clearly affect the dedication and spirit of the team. As the PM, your job is to deal openly and honestly with the issues. It is much worse to see "the elephant in the room" and try to ignore it. When things go badly for the program, your job is to minimize negativity and express (and feel) your confidence in your team's work.

END OF CHAPTER QUESTIONS

For Discussion

1. Procedures and policies can both help and hinder the PM in performing his or her duties. Talk about some examples of this dichotomy.
2. How involved would you want your customer to be? Would your answer be affected by what type of program you are managing?
3. Have you seen programs "go bad"? What is the effect on the people leading and working on the program when that happens? What kind of leadership actions or behaviors turn around a bad situation?

Written Assignments

1. Relate an anecdote of a customer interface that you have participated in or witnessed. What were the dynamics of that interchange? Extra credit for discussing both a positive and a negative experience!
2. As a PM, you have a great idea for the program and tell the team. It does not engender the enthusiasm you expected. Why do you think this might have happened?
3. Related to "For Discussion" question 3, tell us a story about a program that "had turned bad" and how the leadership of the program helped or hurt its recovery.

3

Identifying Opportunities

As a program manager (PM), very often you will participate in deciding the course of your company's pursuits, and this can be in a strategic or tactical way. For example:

- Strategic—Your company is well established in designing and producing military shipborne radar. Is it a good business decision to pursue an adjacent market in commercial ship radar?
- Tactical—Your long-standing customer has asked you to bid on a contract to design an instrumentation system. It will take over 500 hours and roughly $100K to develop a proposal.

In most companies, there are business development departments and senior managers who will be leading/making the decision, but you as PM are likely to have the greatest overall insight into the opportunity.

- Strategic—You will know what the company is good at, how far it can stretch, how cost-competitive it is, and how much technical or manufacturing capacity the company has.
- Tactical—You will know if you have some discriminating technology, if you are currently on the "ins" or "outs" with the customer, how cost-competitive you might be, and who the other likely bidders are (and their strengths and weaknesses).

THE PROGRAM MANAGER'S KNOWLEDGE IS KEY

Although there may be other parts of your organization leading and/ or making the decision as to what to pursue, you may well be the most knowledgeable as to the whole picture. Hopefully, you are recognized as that resource. But of course, people being people, some of your colleagues may be convinced, or at least like to think, that *they* are the most knowledgeable or the wisest. This, of course, can lead to tension. Your challenge is to use this tension to the good of the company by using the different points of view of your colleagues to the best advantage of the company. Even if you feel that their insight is less than yours, open discussion will sharpen your own viewpoint and may even help to improve your own insight.

But how? Here's where the "art" aspect of leadership comes in. If you have been treating people with respect, helping them whenever you can, helping them avoid embarrassment, and, in general, supporting them when they need it, you have built the relationships that will make your insight valued and appreciated. One act of unwarranted aggression toward a colleague ruins the dynamic for years. Avoid getting angry with those who are disagreeable or uncooperative—or if you cannot help getting angry, do not act on it.

Then, having established yourself as a trusted, helpful colleague, not a roadblock or a hothead, you are in a position to use your expertise to evaluate opportunities. Since most of the time PMs become involved in the more tactical kinds of pursuit decisions, we will examine some of the features that should be considered.

Early in my career, one of my friends and colleagues decided to take the half-hour drive to another company location to talk to someone in another department about finding some engineers to work on a new project. We must have caught the fellow at a bad time, because he became angry and proceeded to lecture us on how busy he was and why we should have made an "appointment." While it really was not in our culture to make an appointment for that kind of meeting, we listened to him vent and finally he took the 3 minutes to point us at a collection of résumés that did in fact prove helpful. But what effect do you think the getting angry and lecturing us had on *his* effectiveness? It hurt him disproportionately to the relief he may have felt in venting. Because even though we realized his blowup was due to stress from some other area, we made the conscious decision to avoid him in the future. He took one step further into the intellectual isolation that difficult people often "achieve." This one event, remembered for years, compromised his ability to contribute to the organization. Do not let it happen to you! Figure 3.1 illustrates the encounter.

FIGURE 3.1
Next time, make an appointment!

(From Dollar Photo Club, File #59923953 emrCartoons.)

- Is it a *real* opportunity?—Sometimes customers are sure they want to award a contract to a particular company, but the Federal Acquisition Regulations (FAR) or their interpretation of them may make it necessary to open up the opportunity to competitive bid. So they may just be "going through the motions," knowing that some-one, for example, the incumbent on a particular project, has a dis-tinct advantage and is the likely winner. If, for example, a competitor is favored because of its past good performance, and may have a distinct technical edge, it may be a waste of your company's time and money to attempt to dislodge this competitor. But incumbency does not always mean advantage—if your competitor's performance has been poor, the customer may actually favor a change. As PM in a particular business area, you probably have insight as to whether an opportunity is real or not, and your company is served by that knowledge.
- What are the risks?—The usual primary risk is that you would bid on and win a job and then lose a lot of money on the contract. Depending on the nature of the work, the type of contract (types of contracts are discussed in Chapter 1), and your company's talent, the risks may be manageable. Ideally, your customer would not put out a request for quotation (RFQ) asking for a fixed-price bid on challenging developmental work. But the customer might, and you would be in the best position to detect the inappropriate risk-reward balance. Beyond the financial risks, there are others as well. What if

the work is more difficult than your customer realizes, and, *even in a time and material (T&M) contract,* your company fails (in the customer's eyes) to solve the problem? There is a chance that the ill will generated on the T&M contract could spill over into other dealings with this same customer and adversely affect ongoing work and/or the opportunities for more work. Maybe failure on this contract, or on products it produces, could have an adverse effect on your company's reputation. Or maybe there is a lurking patent infringement that will get your company sued. Now, some of these risks can be negotiated down, perhaps even to zero, in negotiating the terms and conditions of the contract. Seeing the path to an acceptable risk position should certainly be part of the evaluation of any opportunity. Always remember that it is impossible to avoid all risk—except by not taking any contracts, but that is the path to closing your company's doors.

- What is the necessary commitment?—If an opportunity requires that your company's experts be assigned to the contract, and if those experts (as is always the case) are in short supply, is the opportunity then "worth it"? Clearly, if you have five engineers who are power electronics specialists, and a contract would require all of them, you are blocking yourself from other opportunities, at least in the near term, until you can hire more. So if you are going to have to commit scarce and valuable people to a program, it has to have sufficient strategic merit (in establishing relationships and reputation, long-term production, etc.) to make it worthy to pursue.

- Who is the competition?—Very often this is not readily apparent or discernible. Depending on the nature of your business and any FAR rules that apply, you may be able to get some insight and/or make some educated guesses. Often the best approach is to ask the customer questions. He or she will be bound by rules and may not be allowed to answer, but it does not hurt to ask. Very often even the way the question is not answered may give you insight into the nature of the competition. If a customer is anxious for you to bid, it may mean that you are the only viable competitor. Or it may just make the procurement easier for the customer if there are two bidders. If the work is foreign to you, and the incumbent is favored and knowledgeable, then you may be spending money just to make it easier for the customer to select a competitor! The better you know the customer, and the more alert you are in the nuances of even the "no

FIGURE 3.2
Making informed choices.

(From Dollar Photo Club, File #62404776 McCarony.)

comment" answers, the better chance you have at marking a valid interpretation.

- What is the payoff?—Is this pursuit "worth it"? Will the company make a profit that justifies the expense in developing a proposal? What kind profit levels can be achieved? Is this contract the basis for follow-on work that can be won with little expense and at high margins? Or is it more of a dead-end project that is not likely to lead to other work?

Remember that since your company has a limited amount of money to pursue new business, you should see these opportunities as *competing* opportunities. Your insight allows the company to peak behind the door before you select it. Figure 3.2 shows an opportunity to peek in before entering.

PROGRAM MANAGER OPPORTUNITIES

We have spent some time discussing how the PM can help identify the right opportunities for the company, but now we will discuss how the PM can find the right opportunities for him or herself.

Often, the proposal manager on a particular bid becomes the PM. This evolution makes a lot of sense in that by the time the proposal is put together, he or she has the best understanding of the job, the customer, the challenges, the risks, and the opportunities. Therefore, the proposal manager should be a "believer" in the program that is being proposed, and hopefully, senior management recognizes this point and makes it a prerequisite for selecting the proposal manager.

If you have been chosen and accepted and are very positive about the program, then it is probably for you. Sometimes, however, because of one's availability, capability, and/or perceived capability, the proposal manager is not available to manage the program. Thus another PM, who may have had nothing to do with the proposal, is "offered" the opportunity to manage the program. This could be you. The reason that "offered" is in quotes is because at some times and in some cultures, it is not practical to say no—or if you do, there may be some consequences. As alluded to before, however, saying no may be the best thing for the company. A leader who is skeptical about a program's success is going to be a bad leader—if he or she expects failure, ßfailure is more easily accepted. Good leadership selection requires choosing leaders who are anticipating success.

Some years ago, one of my colleagues was selected to lead a program with which he had not been involved. He did not think the funding or schedule that had been agreed to by the proposal team and senior management was achievable, and he said no to the "opportunity." For a few months, he was seen by senior management as being "selective," which was not viewed as an endearing quality. Hence, another PM was found who was not so selective, and she had a terrible three years, trying to do the work that essentially did not fit the money available. While her efforts were appreciated, losing money (even if that loss is minimized) is not career building. Plus, the first guy went on to other projects in which he did believe were outstanding and was very successful and fulfilled. Sometimes it is OK to say no, and though uncomfortable at the time, it may be the best thing for both you and for the company. The first PM had a few difficult months of being labeled "selective" and three years of challenging, positive work. The second PM, though not "selective," suffered 3 years of trying to do the work with insufficient budget. Figure 3.3 shows that you have a choice: when to embrace and when to renege!

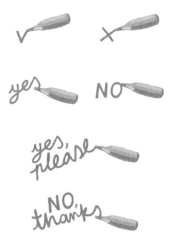

FIGURE 3.3
Knowing when to say "no."

(From Dollar Photo Club, File #51365334 demonique.)

Thus, in summary, as PM you have a lot to contribute to the decisions your company makes in developing new business. To be an effective voice in applying the insight you have, it is important to develop relationships that make folks want to hear your opinion. Good relationships are vital to your success and your company's. Finally, although your ability to choose the best opportunities for yourself may be restricted, it is to your long-term advantage, and the company's, to be "selective."

END OF CHAPTER QUESTIONS

For Discussion

1. What are some reasons that your company might want to bid on a contract on which there is little chance of winning: for example, a production contract where the incumbent seems to be in favor with the customer?
2. Suppose you are offered an opportunity to manage a program that you think has been underquoted. What are some of the factors you might consider regarding whether or not you can accept this opportunity? If you say "yes," what can you do at the outset to make the experience positive for your career growth?

Written Assignments

1. Suppose your company traditionally concentrates on ground-based radar. Discuss the considerations of broadening the company's horizons into an adjacent area (of sorts), for example, airborne radar.

2. Have you seen how an effective PM can take an underfunded program and still make it successful? If you have, tell us about it—if you have not, tell us what a PM might do if he/she finds him or herself in a situation where there is $2M worth of work to do and $1M worth of funding with which to do it.

4

Pre-Proposal Work

Found an opportunity? Great!

In the last chapter, we discussed how the correct opportunities could be selected and how the program manager (PM), perhaps operating for a time as the proposal manager, could help identify opportunities that would be successful. So once the target opportunity has been identified, it's time to relax until the customer's formal request for quotation (RFQ) or request for proposal (RFP) has arrived. Right? No, absolutely wrong.

A contract is likely to be won or lost by what goes on before, even well before, the RFQ or RFP has been issued. Take the following case, for example. Your customer feels that there is a need for increased capability in a system that you have been providing for years. With advances in technology, it is possible to provide that increased performance. Now, depending on how strategically important this business area is to your company, you can do one of three things:

1. Invest in the technology that is important to the customer.
2. Gather competitive information.
3. Wait for the new specification, and bid to it, and hope for the best.

As you can imagine, these three strategies involve different levels of work and investment, and depending on how badly the company wants to win, (1) and (2) may both be used. Generally, (3) is a poor strategy. While it lessens the cost of bidding, it also greatly lowers the chance of winning. Thus, this seemingly inexpensive approach may be just throwing money away. We will discuss the three approaches in turn:

1. <u>Invest</u>. Use discretionary funds (typically called IRAD—internal research and development) to get a jump on the technology—possibly

developing a demonstration of the new technology that will cause your customer to have confidence in your ability to advance the technology and provide the desired improvements. This approach, however, can be very expensive, and it is a bit of a gamble that you have correctly identified the features that are important to the customer. Clearly, the long-term profits that can be expected from this approach would have to be significantly greater than what you invest. However, if funds can be made available, not only do you have the potential of leapfrogging the competition, but also when it comes time to respond to the RFQ, your price will be lower because of the investment already made. In fact, this advantage could be even more substantial for development programs, where (valid) fear of the unknown may force you to include risk dollars as well. A working demonstration removes the risk (and the risk dollars) from your proposal.

2. Gather intelligence. Compile all the intelligence that you can about the forthcoming RFQ, including what is important to the customer, the evaluation criteria that will be used to evaluate competitive proposals, who the competition is likely to be, what the customer's budget allocation is, what kind of schedule might be required, etc. Some of this information may be readily available, but one must be careful not to violate procurement rules or ethical behavior in the quest for information. The best and often the most effective approach is simply to ask the customer (through the proper channels—usually your contracts department to the customer's procurement department) the types of questions that can help you frame your strategy. It can be dangerous to seek this information engineer-to-engineer. If you have a good relationship with the customer's engineering staff, it can be even more dangerous. Most customers' technical people are well versed in keeping privileged information to themselves, but if an inappropriate question is asked and inappropriately answered, it can adversely affect the customer's procurement plans. For example, you would certainly like to know who might be your competition for a particular project, but it is not a good idea to ask the engineer because he or she should not tell you, and if he or she slips and does tell you, it would put him or her and you in a compromised position. In the worst case, you might be excluded from bidding, or your customer

may have to cancel the procurement and start again. This is not the way to win programs. Figure 4.1 illustrates the wrong way to gather information.

There are, of course, legal and ethical ways to gather information. Often you can make an educated guess as to who the competitors are. You can take advantage of published information about the company, perhaps disclosures to Wall Street that may indicate how aggressively those competitors may be working toward this procurement. Or perhaps you have had experience competing with this contractor before. If you have been unfortunate enough to have lost to that competitor, you would be entitled to losing bidders' information. Typically, losing bidders are told how many bids were solicited and how many bids were received. And, most importantly, what the dollar value of the winning bid was.

FIGURE 4.1
Gathering intelligence the wrong way.

(From Dollar Photo Club, File #34532241 G. Nicolson.)

One of the most important legal pieces of information that you would like to have is typically available to you. Most government procurements have evaluation criteria to make the process as equitable and auditable as possible. Points are assigned to each facet of the proposal, thus developing a weighting system. One example might be as follows:

a. Technical approach: 60 points
b. Price: 30 points
c. Past performance: 10 points

Knowing the evaluation criteria is extremely helpful. In this example, you should spend your proposal money making sure that your technical approach is sound and clearly explained to the customer. Price is important, but a poor score on the technical approach can easily cause you to lose. Of course, that does not mean you can ignore price—after all, if your competitors' technical solutions are considered to be equal to yours, then it will all come down to price. (There is nothing you can do now about your past performance record—hopefully, your company met all of its schedule and cost targets!) Losing bidders' information from past procurements can help you here also. Very often, the customer will also tell the losing bidders how they scored in each of the categories, and may even provide some information on how the winner scored.

Also important are the rates (hourly charges) that your competitors charge. These numbers, of course, are highly guarded and highly proprietary. There are, however, ways to get valid estimates of them. Past proposals in the public record are the best source. Very often your marketing department can do an analysis for you, or they can hire consultants who will (legally) develop information from publically available sources. Figure 4.2 illustrates the work required to

FIGURE 4.2
Gathering intelligence the right way.

(From Dollar Photo Club, File #63167410 Leremy.)

gather the information the right way. If it is decided to bid, there may be a formal "Black Hat" process that specifically addresses known or suspected competitors' capabilities.

3. <u>Bid to the specification, and hope for the best</u>. As one of my VPs once told me, "Hope is not a strategy." And he was right. Remember, you are likely competing against companies that have done their homework and are already using investment and aggressive information gathering to beat you. What is the likelihood that without those efforts you would be able to develop a competitive bid? In fact, if you shortchange the proposal process, you could possibly develop a price that is too low, and win, and be very, very sorry that you did. It has been said that "No business is better than bad business," and if you have the misfortune of managing a program that has been underbid, you would definitely agree! If your company is only moderately interested in the particular contract, it may be tempting to do a superficial job on the proposal, minimizing costs, just to gather the losing bidders' information for the next competitive procurement. You may want to think twice about doing so. A lackluster technical proposal will cause you to lose points with the customer. And if you spend too little effort evaluating the price, you may just win it at too low a value, producing years of "recovery plans," "extra help," and sleepless nights. It is far better to decide what bids you really want to win and spend the necessary time and money to pursue them adequately.

USING PRE-PROPOSAL EFFORTS TO DEVELOP A WINNING PROPOSAL

Hopefully you are now convinced that the winning bid has its origin in pre-proposal work. Often, using the usual paradigm of the proposal manager becoming the PM, you will be asked to develop a budget for the pre-proposal work. As alluded to above, specific pre-proposal work could be very modest and be completed in a month, or for especially strategic wins, may actually be a mini-program in itself. You will likely be asked to develop that budget, explain why your proposed plan is the right approach, and—then you will be asked to do it for less! This is not an uncommon approach. It might be called "a challenge" and indeed it is. Knowing how hard to fight for your budget, and what strategies you might use, are all very dependent on your company's culture (and available funds). The key theme, of course, that you will want to present is this: the more money that is allocated for pre-proposal work, the better the chance of winning. In fact, it is generally accepted that money spent early in the procurement cycle is much more effective than money spent in developing the actual proposal. This makes supreme sense when one realizes that in the preproposal phase, the proposal strategy is developed—and the staff at that time might only be one or two people. When the proposal effort starts in earnest, there might be 10 or more people executing that strategy. It certainly makes more sense, efficiency-wise, to be sure that the strategy is sound (even if it takes a few dollars) than to start off with a strategy that is only inadequately planned and to waste effort working to an incomplete plan.

So what should the proposal manager do with the funding? Certainly, some of it should be devoted to intelligence gathering as described earlier. As for technical investing, it is important to have a good understanding of what the customer considers to be important. If your current radar can track eight targets simultaneously, and it will cost dearly to raise the performance to track ten, it is very important (and obvious) to know if the customer needs that feature. The time to gather this information is way before the procurement is imminent. Technical conversations with your customers' engineers are invaluable in determining what is important to him or her. When there is no procurement open, or even planned, your customer can be (and should be) more open to discussion. It is therefore important that your technical staff is alert to this intelligence and organizes and uses it with you. A good practical way to do this is to get your technical folks together and compare notes. You might put yourself in your customer's place, trying to understand his or her thinking. So maybe you do not want to spend your money improving your radar to track 10 targets—maybe the customer has a weight problem on his next platform and saving 20 pounds is what is important. Technical personnel in your company, who are alert to customer needs, are the key to refining your approach to be what the customer wants. Providing extra features to a customer that wants to simplify the user interface is more counterproductive than positive.

As you strategize, remember the customer is not a monolith. What might be important to Customer Engineer A might be the opposite of what is important to Customer Engineer B. Thus, inferring the best approach brings up a second issue—is Customer Engineer A or Customer Engineer B more influential and more in tune with the broadly perceived issues?

Comparable to most of your intelligence, knowing who is influential relies on good communication with your customer. If you have worked with both Customer Engineers A and B, and worked with their colleagues, it is likely that you will have a sense of which approach is favored by the customer's decision makers. Speaking of that, would it definitely be good to know what is important to them? You bet—but they are more distant from you—both administratively and also, in many cases, for reasons of levelism. Their VPs may only want to speak to your VPs. If this is the culture, it is your job to involve your VPs in the process. They have to be briefed, perhaps, on what kind of information you might want, but rapport at that level, which you have hopefully built at your level, is obviously important to identifying where to concentrate on your proposal.

OTHER CONSIDERATIONS

If you are evaluating an opportunity with an existing customer, you are likely to have a good grasp of the working relationships between you and them. But with new customers, it is important to understand their expectations and protocol. Many of the hard and fast rules are spelled out in the FAR (Federal Acquisition Regulation) for direct government procurements, but even so, there are expectations and traditions that the customer has that you would want to use in developing your proposal strategy. Many times you may not be bidding directly to the government. You may be bidding to a procurement agency or to a prime contractor. Here, the FAR requirements are less clear, and knowing the customer is even more important. Very often, the procurement will be preceded by an RFI (request for information). By using this vehicle, the customer may be trying to find out what technology is available and what companies might be good candidates to solicit for a bid. This less formal time is ideal for learning what is important to the customer. You should take every opportunity to meet with the customer's representatives both to learn what is key and to impress these folks with the capabilities your company has to meet their needs. Remember, it costs your customer time and money to evaluate each proposal that is received, so the customer will want to narrow the bidding to a manageable number of bidders. Hence, the RFI process is important, especially for customers who may not know much about your capabilities.

Through the RFI and various other processes, the customer will develop a bidder's list, and send the formal RFQ to those on the list. If you are not on the list, it's likely you will not even know.

This is the major way that the government and its procurement agents control the process. It is critical that the customer *wants* you to be the successful bidder. If the customer does not "like" you it is very easy to leave you off the list. Now, this "like" is not as arbitrary as it sounds—again, because your customer is not a monolith. When creating the bidder's list, it is in the government's best interest to have a large number of qualified bidders. So it is not likely that you will be left off because you stepped (literally) on one of your customer's key personnel's toes, or spilled coffee on his or her suit. But, if your company has been performing poorly on a contract with the same customer, or if you have been giving the customer's representatives undue aggravation on terms and conditions or profit level expectations, well then, you can work yourself off the bidder's list, whether you find out about it or not. Remember, there are often more qualified bidders than the customer has the time, personnel, and funding to evaluate, and it is very reasonable to limit the bidders. Your company must absolutely perform well on contracts with this customer, or you will not win the next contract, because you will not even be asked to bid. Put yourself in your customer's shoes—we all have had experience with our own "contracts." You would not ask a plumber to bid on some work for you if the last time he came he caused a new leak. You would not go again to a dry cleaner who ruined your favorite suit or dress. The government procurement process may be a little more rigid and formalized, but you can be sure that the people who do not <u>want</u> to deal with you probably can make sure that they do not have to. You (your company) have to satisfy the customer—<u>continually</u>—on all your work.

SUMMARY

The pre-proposal phase is extremely important to developing a winning proposal. We discussed rules and strategies on how to collect information properly and effectively. We considered the risks of presenting a bid with insufficient work behind it, and we reiterated something that cannot be said often enough: you must continue to please your customer if you want more business from that customer!

END OF CHAPTER QUESTIONS

For Discussion

1. Assume you are the PM on a long-term production program. Your customer has mentioned an interest in increased functionality. Your management might like to keep things as they are on this lucrative program, but you are concerned that your customer might start to look elsewhere. Discuss this dynamic and describe how you might handle this situation.

2. You have a personal friend at a competitor, and you discover that you are both bidding on the same opportunity for a Department of Defense customer. Can you exchange ideas and information in the hope that you can do a better job for your company or the government? Should you disclose this friendship to your company?

3. Your long-term customer is about to issue an RFQ on a new program that would be attractive to your company. How much information can you seek on that opportunity from your customer-friend? Should you involve your contracts or legal department—and if so when?

Written Assignments

1. You have found an opportunity for long-term growth on a budding program that is aligned with your company's capability but goes beyond current technology. Describe the considerations that should be made before you might recommend a large R&D (research and development) project to become proficient in that new technology.

2. You have identified a new potential customer who does not know your company's capabilities. What actions can you take to win him or her over?

5

The Proposal Process for a "Typical Program"

Immediately, we have a problem: there is no typical program. So, what we will do is discuss an example program. We will choose an example that can be applied to both design and production content so as to examine both major types of programs that you may encounter. Of course there are many more types of programs, all with their own particular issues; however, the design and fabrication of complex equipment has enough issues itself to be worthy of special consideration. Whether the anticipated contract is fixed price or cost plus, or time and material (T&M) much of the same process applies. While the risk of losing money is less, of course, on cost-plus contracts, failure to provide an accurate cost picture to your customer at the time of the proposal is an invitation to later problems and dissatisfaction with your company.

THE IMPORTANT PRE-PROPOSAL PERIOD

As discussed in Chapter 4, it is usually unwise to think that the proposal work begins with the receipt of an RFQ (request for quotation) or RFP (request for proposal). Expect to do a lot of work before it is issued by the customer. And, as also discussed in Chapter 4, oftentimes the opportunity is with a traditional customer with whom your company has done business for several, if not many, years. Using your knowledge of what the customer's priorities and issues are is a great head start on creating a winning proposal.

Since it is likely that your company and you have established relationships with your customer—ideally at multiple levels—you should have a good idea that an RFQ is imminent and what its content will be.

Remember that sometimes there will be restrictions on what the customer may tell you—and of course you must respect this. It will do no one any good if you get advance information that you should not have. First and foremost, it is an ethics violation and your company needs to maintain the highest standards of conduct, and will insist that you do also. Furthermore, should your customer "leak" some information to you, he or she would personally face liability, and the procurement process may need to be terminated and be restarted by the customer using different personnel. This, obviously, is no way to ingratiate you or your company to your customer.

So, keeping in mind that some information may not be allowed to be given, it is to your advantage to get all the information that you can legitimately have. The best way is to ask—but because this is a potentially sensitive area, it may be best to ask formally, or if informally, to have your contracts representative present or included in the call if it is a phone conversation. Likely, your contact will be a technical lead person, but if she or he is at all unsure of what can be divulged, you should make it comfortable to include the business arm—usually a procurement specialist or buyer. Similar to your contracts representative, it is that person's job to know the rules and protect the integrity of the procurement process.

TO BID OR NOT TO BID

Remember, the idea is to get as ready for the RFQ as you can before it is issued. This will obviously require funding; in most companies, the decision to bid or not bid is made at a couple of levels (at least) up the organizational chain. It is likely that the program manager (PM) (and at this time the future PM would most likely be acting as the proposal manager) will need to prepare a formal recommendation as to whether the company should or should not bid on the forthcoming RFQ. "Bid and proposal" funding, and proposal personnel, is obviously limited and senior management needs to decide on whether the opportunity is in the company's strategic direction, if the potential returns on a successful bid justify the proposal costs, if the probability of winning is high enough, and if the necessary resources (personnel, plant floor space, etc.) are available. Importantly, you will likely have to identify risks involved in executing the program: can the work be done for the dollars bid, is the technology needed readily available, is there any legal liability in performing the work, are resources available, and will the reputation of the company be hurt if things go particularly poorly. It is likely that you, because of the pre-proposal work you have done and because you have decided it is worth presenting, will be advocating a "bid" decision.

Your senior managers know this and will be, therefore, expectedly skeptical. You may feel they are accusing you of stacking the deck in favor of the "bid" decision—and, even inadvertently, you may have done so. Therefore, you can expect to be challenged, and the better prepared you are in understanding their concerns, the more likely you will be successful in getting the decision that you want. It is always a good interpersonal strategy to "put yourself in your audience's shoes" and try to think like them. You can even rehearse your presentation with a friend/colleague and get some practice answering the questions you may be asked by the VP. (VPs like crisp, accurate answers—I guess they learn this in VP school.) Hemming and hawing makes them nervous, and nervous VPs are more likely to say "no."

Some RFQs will include the opportunity to attend a bidders' conference. This is an opportunity for the customer to ensure that the bidders have the maximum understanding of what is needed and wanted. By all means you should attend, because it is an opportunity to see who else might be bidding and to use that knowledge in your proposal strategy. At these meetings, you can expect a certain amount of "tight-lippedness" among the potential bidders. While it is an opportunity to impress the customer about what your company knows and can do, it is also a (negative) opportunity to divulge this information to your competitors. So, they know this too, and these meetings are very often one-way, and any questions are fairly innocuous, such as, "How many copies of the proposal would you like to have?"

As you prepare to bid, do not forget to seek out the experiences, both positive and negative from prior similar proposals to this customer. The negative ones will most likely yield the most knowledge. Any unsuccessful offerings will most likely have had "losing bidders' information" and, ideally, a losing bidders' conference. Make sure you see the people who were involved in those proposals and find out what they wished they had known before the bid was submitted (or even created). (See a more detailed discussion in the Section "Reviews.")

DEVELOPING THE STRATEGY—GETTING STARTED

Assume you do get permission to propose and are given a budget to create a proposal. Now the real work begins. Very often you may already be working on a similar program, and this proposal will be extra work. The same applies to the people that will be working for you on the proposal. Remember that even with all the intelligence you may have regarding an upcoming bid, when the RFP or RFQ actually comes in and when you actually get permission and funding are unpredictable. Thus, preparing a proposal is often very demanding on your time, and, depending on your seniority in the company, its rules, and its traditions, you may or may not be compensated for the extra work. Getting to frame the job that is likely to be yours may be considered to be adequate compensation in itself.

So, having successfully convinced your superiors that this is a bid that should be undertaken, you will be in an excellent position to apply your arguments to the start of a successful proposal. In your presentations to senior management, you would have had to describe why your company could win. This may include things such as:

- You know that your company has superior technology (including patents and trade secrets).
- You know that your hourly rates are more favorable than your competition.
- You may have transferable work products from other projects (such as reusable software).
- You own capital equipment (for example, surface-mount printed circuit board assembly lines) that allow you to do work in-house for which your competitor would have to contract outside.
  - Remember that when you subcontract for work, you are immediately at a potential cost disadvantage. You must pay your subcontractor its cost and profit, and then apply your own general and administrative (G&A) markup and your own profit. Thus, subcontracting makes sense only for goods and services that you cannot accomplish economically in-house. (And high internal costs are a poor reason to need to compete with yourself!)
- You may know that your customer favors your organization for some tangible or intangible reason—usually successful past performance on similar projects.

In most large procurements, selection criteria will be provided. It might allocate points for the following:

- Technical approach or technical content
- Program management experience or confidence
- Prior performance
- Price
- Low schedule risk
- Low technical risk
- Use of small or disadvantaged businesses

Using your knowledge of the procurement content, the customer, and the scoring criteria discussed earlier, it will be your job to develop a "win strategy." Succinctly put, "How shall we create our proposal to give us the best chance of winning the contract?" This strategy has both financial and nonfinancial content. Some of the selection criteria may seem to be out of your control. For example, prior performance: If you have had ten similar contracts with this customer and eight of them have been excellent, and two of them a disaster—then the facts speak for themselves, right? No, wrong. Your proposal should emphasize why this project is similar to the eight that were successful and not like the two that were not. If you can convince the customer that the basis to evaluate "prior performance" should be the eight successful projects, you can get all the points for that criterion.

By honestly evaluating your company's strengths and weaknesses against the selection criteria, you can see what aspects to tout in your proposal and what to de-emphasize or minimize. But writing a winning proposal is likely to require more than just clever writing. To win, you may also have to adjust how you do business—for example, if price is important (and it almost always is), you may have to devise a way to lower your costs—allocate capital money to make fabrication costs less, select personnel for the job who are as low cost as possible—that is, quote more junior (read "less expensive") engineers—use contract labor, which may have less overhead costs than your in-house labor for some work, make "make or buy" decisions that may be internally unpopular but provide a cost advantage, or insist that your procurement department take a 10% challenge (and ensure they meet it!).

Winning the contract involves not only how to write the best proposal but also how to arrange the performance approach that will yield the best pricing. You should "own" this win strategy, but it does not mean that you have to develop it by yourself. In fact, that is not a good idea for a couple of reasons:

- No matter how smart you are, you are not as smart as you plus another few smart people.
- If your approach requires buy-in from others in your company, the best way to ensure that buy-in is to have those others be part of the strategy development. (Remember, everyone—including you—has the most affinity for his or her own ideas!)

"PRICE TO WIN"

As we said earlier, price is virtually guaranteed to be a key, if not *the* key aspect of your proposal. Thus, oftentimes, it is recommended or required (sometimes by senior management) that a "price to win" (PTW) be calculated. Very often, the PTW will be performed not by the proposal team but by a separate individual or team. And sometimes the PTW can do more harm than good. Senior management is likely to put a lot of stock in it, and that is the risk. The PTW calculation really is not a "calculation" at all. It is an educated guess.

Very often the, marketing department is responsible for the PTW. Using publicly available information (for example, a customer's labor rates as divulged in similar, past contracts), and what is known of a competitor's strategies are useful. Marketing may develop a PTW that has five significant figures! Since the process *seems* so exact, the PTW can take on an unreasonable prominence in the proposal process. Set too high, an unwarranted complacency on how to provide the lowest cost proposal may set in. Set too low, internal, high risk, cost-cutting strategies may be put in place, and you could win a contract with insufficient funds to execute it. (Winning a contract at too low a price will cause pain for the entire duration of the program—and is really, really a bad idea!) As proposal manager, you need to keep the PTW in perspective in your own mind and in the minds of senior managers. You need to evaluate just how much guesswork was involved in its creation and must scale the reliance upon it accordingly—both in your mind and in the minds of your managers. Many proposals have met the PTW, and were lost, nonetheless, on price!

LEADING THE PROPOSAL TEAM

Remember, a proposal effort is much like a mini program—it has a product (the proposal itself) and is created by a team of dedicated people. Your job as proposal manager is to lead this group of dedicated people in the creation of a successful proposal. The work involved in doing so is much like the work that the PM will have to do if the proposal is successful. Thus, success in the proposal establishes excellent credentials for the proposal manager to become the PM.

After establishing the proposal development budget, the primary tasks are to develop a work breakdown structure (WBS) and the proposal schedule. There will, no doubt, be a proposal due date, which is absolutely sacrosanct. If you miss that date, your bid will not be considered, and you and your team's work will be in vain. Therefore, the development of (and adherence to) a reasonable schedule is of utmost importance. One common error is to develop a plan that does not include adequate review and recovery time. Figure 5.1 is a typical (simplified) schedule that illustrates this need. You can see that almost half the time of the proposal development needs to be allocated to the review of the work and incorporation of the findings. (The nature of each review and its purpose will be discussed later.)

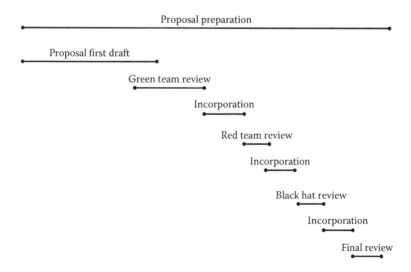

FIGURE 5.1
Example top-level proposal schedule (emphasizing reviews and recovery times).

Next, you must establish a vision for the proposal. What form will it take? What will be the theme—great applicable experience, technical know-how, robust production capability, success on similar programs, etc. Will there be extensive illustrations? Examples? Diagrams? Will there need to be artist-quality diagrams?

Very often the RFQ instructions will dictate things such as—separate volumes for program management, technical offering, and pricing. Your company may have some "stock" proposal boilerplate that talks about facilities, capabilities, and the like. This can be a valuable resource in the PM volume. (Be sure to read it in detail; it may have residual material aimed at a different customer—or worse, mention that customer by name!)

BUILDING THE TEAM

Concurrent with the establishment of the vision of the proposal, you will be working on establishing your program team. Depending on your company's size, culture, workload, etc., you as proposal manager may have varying degrees of control of who is on your proposal team. Do your very best to get the most talented people on the job and check that they have mutual respect and can work together. As with any project, its success depends largely on the people assigned to it—and since proposals are key to the growth and even the survival of the company, you should be able to command the very best personnel. (Truth be told, some companies have an "A" team just for proposal work. But this is a potential risk—customers working with the "A" team in the proposal phase are sometimes dismayed that those folks have gone onto the next proposal and new folks, possibly not as talented, will be working on their newly awarded program. This approach has the potential of looking like "bait and switch" and puts the execution team in the hole from the beginning. It is a better practice, for many reasons, to have the key proposal personnel, in particular the proposal manager, be principals in the execution of the program as well.)

Your team will help with establishing the proposal vision—and the more they "own" the vision with you, the better the quality of their work.

This is due to the "passion factor" inherent in executing one's own ideas versus those foisted upon them. One of the most important principles of effective leadership is allowing your team to execute its own ideas. And like any other team activity, it is important to make clear assignments and divisions of responsibility. That is not to say that folks must work in silos and not cross-pollinate ideas—far from it. But it must be clear who is responsible for what—otherwise at the end you find large gaps in your vision!

Part of your strategy will be to determine, with the help of your team, how compliant your proposal shall be to the RFQ requirements. This is often a serious challenge. Sometimes the customer may ask for performance, system features, or topology that you do not agree with from a technical perspective. Or the customer might impose quality or second source requirements that will make the procurement unreasonably expensive. Or administratively, some of the liability clauses and/or payment clauses may be very unfavorable. You can take exception to these requirements in your proposal and clearly state those views. But that is not without risk. Very often the evaluation team would have to ascribe a dollar value to your exception, and since you are unlikely to know what was ascribed, you could be putting your offering in an unknowable jeopardized position. And, if the exception is in an area that is an absolute to your customer, your proposal may actually be set aside for nonconformance.

Often, it is in your best interests, and in the customer's, to have a dialogue on these issues. Often, also, your exceptions and clarifications will be shared with your competitors. There is a risk here, of course, also, in that your exception may give your competitors insight into your strategy or your company's capability. If it looks like you have a weakness, they can "ghost" your weakness and say in their proposal how particularly good their company is in that area.

Thus, there is a lot to be said for taking as few exceptions as possible, but there is even more to be said for using appropriate exceptions as an opportunity—even if they are open to your competitors—to get closer to the customer and to learn through this process what is important to them. Be sure in this process, to use any information about your competitors that you may glean from reviewing their exceptions and clarifications that may be disclosed to you through this process.

PROPOSAL PREPARATION—STORYBOARDING AND TEAM DYNAMICS

One effective way to ensure that your team is all moving in the same direction is through "storyboarding" the proposal. Although seemingly archaic in this electronic age, storyboarding is still an effective and widespread approach to proposal creation. In general, the process is to envision the proposal as a story, and create poster board "scenes" in the story, just as might be done in the motion picture industry. (Depending on the idea, sketches or pictures may be more effective than words—or they can be distracting—it depends.) These panels are placed on the wall of a conference room (a "war room") where much of the proposal team will be working for the duration. Even if you do not have the luxury of the entire proposal team being in the same location, a "war room" offers the advantage of collecting the information in one area and providing a visual status of the proposal. Each board will communicate the vision of that section of the proposal, and because it is posted, it will allow the team to add ideas and suggestions as they work on their own assignments. This collaboration will yield many rich ideas and build team ownership for the project. It is an easy way to be sure that the proposal's themes are interwoven into the proposal, which will promote the customer-reader's buy-in. It is also a great way to track the progress of the team and ensure assignments are understood and embraced. A brief daily meeting is recommended so that folks can describe their progress and seek help if needed. You can provide guidance and share the results of your ongoing efforts to assess progress and cost management services for the team.

Also, this process is conducive to identifying and leveraging weaknesses that your team may know about your competitors. While it is at minimum rude, and at maximum libelous, to say about a competitor something such as, "Company A is known to be weak in system engineering," you can certainly say how strong your company is in system engineering. Customers will connect the dots for you—since they probably know something about your competitors' weaknesses anyway; they will be pleased to hear that your company sees itself as strong in that area. This is called "ghosting" the competition and, if used effectively, can raise your company's image by contrasting your strengths against competitors weak in the area. If indeed your competitor is weak in an important aspect, your strength in that area can be raised to the level of a theme of your proposal, getting it posted on the appropriate storyboards, and then worked into the text in several places in your proposal.

PRICING

While the proposal text is being generated, you will no doubt be leading another critical effort—establishing the proposed price for the contract. You will need to prepare a detailed "statement of work" (SOW)—or in some RFPs it will be provided for you. This defines exactly what must be done to fulfill the contract. In most organizations, this SOW becomes what each of the functional organizations quote to. The concept is that the SOW defines the work so clearly that each department knows exactly what has to be done and can therefore provide an accurate estimate. These functional estimates (engineering, operations, materials, test, quality, etc.) are then added up, priced using current rates, a profit margin is added, and, voila, you have the proposed price. And if the contract is won, each of the departments is given an allocation (budget) to do the work that they quoted.

In theory this is very simple, but in practice, it is very far from simple. Your best efforts to write a clear SOW will still nonetheless leave much to speculation, and the functional departments can use it to excuse themselves for inaccurate estimates, untimely responses, and if the bid is successful, for lesser performance ("that was not clear in the SOW"). (These excuses can be either conscious or unconscious—and it may not matter too much—what matters is making the SOW as clear as possible.) Your job as leader (rather than "manager") of this process is to forestall negative behavior. How would you do this?

These are examples of a "blame culture" wherein everyone is making sure that he or she is "covered." As a leader, you must break through this defensive behavior with all its negatives, and create a "we are all in this together" environment. Depending on your company's culture, and the individual character of the people with whom you work on the project, this transformation of local culture will either be simple or very hard. This is another reason that you want the best people on your project—stay away from the defensive, confrontational types and welcome the collaborative, team-oriented types. Ironically, many of these cultures were established in the "old days" when it was assumed that the interactions would be adversarial, and the controls would be punitive. Most successful companies are migrating away from the blame culture, but its legacy does live on—especially in times of stress. All you can do as a leader is build a supportive, collaborative local culture by your own behavior and by selecting team members who are more interested in successful collaboration than in making themselves look good at the expense of other team members. (This toxic behavior must be rooted out from your team whenever it is encountered.)

As the price is developed, it is important that this team ownership continues. Since the effort that is quoted will, if successful, become the allocation to do the work, it is human nature to make it comfortable for oneself. Countervailing this is the realization that if everyone quotes comfortably high, then it is likely that your company will not win the contract. But there is always the chance that, *unless you establish a true team environment,* each department will bid high, hoping that other departments will bid aggressively. You must lead the team through this "selfish self-preservation," and get fair bids from your functional department colleagues. When this is done, it is now time to make sure that senior management does not burden the proposal with unreasonably high profit percentages. Sure, we all want to make money, but too much profit applied to a relatively comfortable estimate probably leads to losing the job.

PRICING STRATEGIES AND RISK MANAGEMENT

Remember that your bid is a guess as to what it will take to perform the contract, and as such it is subject to error. Furthermore, on longer programs, there may be opportunities to add scope at correct prices, and of course there are always risks to lose money! Identifying these risks and opportunities are part of your job, and in many companies are formally tabulated and analyzed. If you put a dollar value and probability on each risk and opportunity, your decision makers can decide how aggressively they want to pursue the contract. A low-risk job, with many possibilities for additional work and/or money, will be much more attractive to senior management and may lead to more aggressive pricing—i.e., a lower profit percentage added to the quote. On the other hand, since losing money on a contract is never, ever a good idea, risky programs with little chance for positive opportunity will be ascribed higher profits, to cover those risks and to make the stress worthwhile!

In Department of Defense contracts, oftentimes FAR (Federal Acquisition Regulation) applies. In those cases, it is illegal to misrepresent costs for the purpose of winning proposals, and strategies such as "we will make it up on changes" are completely disallowed. Remember, honesty is not only the best policy, it is the only policy; and as tempting as it may be to use a "trick" to win a contract, winning is less important than integrity (or jail)!

REVIEWS

Assume you have gotten through all of these challenges and you have a draft proposal. Now it is time to take that draft proposal and make it a finished product. This is done through a series of technical and management reviews.

The first review that you are likely to have is a Green Team review. This is a review by colleagues and advisors who have not been part of the proposal development process. Ideally, you would have a say in selecting who gets to help you—the more you respect these folks, the more likely you are to value their comments. Expect them to review in a collaborative, congenial, and helpful way. In fact, they may even make contributions to the proposal itself. But like all the reviews that your proposal will have, there will be findings and discoveries that will need to be addressed. As the proposal manager, your word is final, but you would be foolish to ignore the suggestions that they make. Once you review and evaluate the comments, you will need to lead your team in the incorporation of the comments. This may be another leadership challenge—your team may not agree with the comments, but your job is to understand their counter-opinions and adjudicate a resolution. As a last resort, you may have to remind them of your position of power—"I AM THE PROPOSAL MANAGER"—but do all that you can to avoid this type of direction—think about how you would feel as a team member and what that kind of "leadership" would do to your ownership of and enthusiasm for the proposal!

When the Green Team's comments are incorporated, it is time to move on to the next review: the Red Team. This review is intended to be a little more adversarial in content. Its members are likely to be more senior and selected by your senior management. Expect "no-holds-barred" comments. Often consultants from outside your company may be invited to be part of this review, so there is an expectation of future collaboration to soften their observations. When you get the Red Team's comments, you can expect even more resistance from your team. You are, after all, the proposal manager and directly responsible for the proposal quality (and to some degree, its results!). Because the Red Team may have been chosen by your manager, your ability to waive their comments will probably be reduced. Disagreements here will provide yet another chance at leadership—in this case, upward leadership!

After the Red Team's comments are incorporated, you may feel the proposal is ready for submittal—but often, you may have one more review to go—a Black Hat review. This review again brings in proposal outsiders who review the proposal from a competitive/customer point of view. (Sometimes, the Black Hat review may be done as part of the decision-to-bid phase of a proposal. When and how are subject to the company's expectations, the situation at hand, and the proposal manager's recommendation.) Ideally, they will bring knowledge of the strengths and weaknesses of you competitor(s) and a circumspect understanding of your organization as well. Expect advice on how to adjust your proposal based on this knowledge. This review may even affect your win strategy. If it does take place near the end of the proposal development cycle, any change in approach is likely to cause some hurried work—and maybe some late nights for you and your team. C'est la vie, as they say in France.

Hopefully, you now have an appreciation for the schedule outlined in Figure 5.1. The reviews themselves and the incorporation of the results of the review are inherently time consuming. You must plan from the beginning for them; otherwise the value of the review will be lost in a torrent of rush and confusion.

And regardless of how hurried the endgame gets, all of your finalized proposal text should be read over carefully before sending in the proposal. Editing errors can creep in anywhere in the process. Do not assume that any of your team reviewers will read every word. If you are not going to perform this final review yourself, be sure you assign it to someone who understands the mission (100% reading) and has the language skills to identify and fix grammar and punctuation errors.

Early in my career, I was responsible for a part of a technical volume for a computer-based control system. The system needed to be highly reliable, and the design included a great deal of redundancy. One section of the text went on and on, extolling the built-in redundancy of the equipment.

Ironically, the final typist, perhaps tricked by the overuse of the word "redundant," inadvertently duplicated a long descriptive paragraph. As I proofread it, I found it amusing that our text was redundant in describing the redundancy, so I circled the error and wrote an editor's note in the margin: "Even our words are redundant!" Well, the typist, not getting the humor of the situation, left in the duplicated paragraph, and, gasp, added the "humorous" note as additional text: "Even our words are redundant!" Feeling I had caught the error and sent clear instructions to the typist, and since it was up against the deadline for mailing, the volume was printed and submitted without that all-important final read. Only after glancing through the text after the proposal was on its way did I see the error. Argh! Well, we didn't win the contract, but, thankfully, it was because our price was too high, not because I failed to do that all-important final read! (It wasn't good to lose, but I was greatly relieved that my error hadn't been part of the cause!) Figure 5.2 illustrates the unfortunate proposal manager who fails to read the final copy of the proposal!

Developing the proposal, as you can see, is a complex and lengthy process. And of course, this process would be scaled for the size of the proposal. A multimillion dollar proposal will obviously need and deserve more work than a $100,000 one, but the principles are identical. In general,

FIGURE 5.2
The importance of final proofreading!

(From Dollar Photo Club, File #35726945 CMonnay.)

you will want to marshal the best folks to help you prepare the proposal, the best and most trusted advisors to help you refine it, the most critical and demanding challengers, and the most insightful and knowledgeable customer-oriented reviewers. For small-scale proposals, the number of people directly involved might be as few as three or four, but for major proposals the total number of people may be in the hundreds. Your efforts will most likely remain somewhere between these two extremes—but regardless of size, your leadership and contagious enthusiasm are likely to make the difference between success and failure.

BUSINESS REVIEWS—THE SIGN-OFF PROCESS

As you finalize the proposal text, another set of reviews needs to be done. Before your company can send out a formal proposal, you must have the company's commitment to perform the contract if awarded. This is usually a formal process and may involve senior management. It is often called a sign-off. Your company will most likely have established processes that are proportional to the dollars to be signed off and/or the perceived risks to the company—contractual liabilities, third-party liabilities (what if your equipment causes damage or injures someone), loss of reputation, etc. Terms and conditions (Ts and Cs) of the proposal, oftentimes dictated by the customer and reviewed and possibly negotiated by your team, need to be analyzed and accepted. Your contracts and/or legal department(s) may lead this review, but similar to other parts of the proposal, it will fall to you to resolve issues that will likely arise.

Usually, by the time of the sign-off, you have put your own and your team's blood, sweat, and tears into the proposal. You and your team are anxious to submit it. You have done all you can to protect the company, manage your customer's expectations, and develop excellent plans and strategies—strategies regarding not only how to win the work but also how to perform the work when you win. You have reviewed the RFQ, and taken the necessary exceptions and made the necessary clarifications. You believe you are 100% "done."

So, your management knows this and may be concerned that your enthusiasm may cloud your judgment of risk! Thus, part of the sign-off process is for the senior managers, whose signatures are necessary for you to proceed, to become sure that there are no hidden risks or risks that are incorrectly minimized by your enthusiasm. This is especially true in large companies that have a lot at stake, reputation-wise, and may have had recent, unfortunate experiences in spending a lot of their own money to solve issues on similar projects. No one, including you, wants to propose work that will do harm to the company. You may hear from time to time, "No work is better than bad work," because that is true; you can expect close scrutiny before you get the necessary permission to submit the proposal.

In some companies, and dependent on the dollars or risks involved, you may expect multiple sign-offs. You will most likely be the presenter at the first level or two, but if it goes higher, your manager or your manager's manager may make the presentation to the next level up. These presentations at high levels are a source of great tension—your superiors do not want to present a proposal for sign-off that has errors or omissions. So expect even more scrutiny.

Just like the need to get the first draft of the proposal done early to allow for reviews and improvements, the same strategy needs to be applied to the sign-off process. Typically, the sign-offs are "orchestrated" by the contracts or business management departments, but you can be sure that you will be responsible for much of the material and, as always, for the successful submittal of the proposal after clearing these hurdles. Keep in mind that your company's senior managers also want to get the proposal out on time and want to win the contract. But they do see themselves as the "final filter," ensuring that no proposal is submitted and accepted that will cause years of angst and suffering for all those involved. They may fear (and possibly because of some recent experience) the VP saying: "Who the heck signed off on this contract anyway!" None of your managers want to raise their hands in this scenario, so you can be sure that caution will be the watchword at sign-off time!

NONCOMPETITIVE PROPOSALS

The process described above is characteristic of competitive proposals, ones in which a customer is formally deciding which company can best, and most affordably, perform the work that it needs done.

In contracting space, there are, however, a number of noncompetitive opportunities that your company may be asked to bid on. In these cases, the customer has decided that there is some compelling reason to only use your company. This, of course, is a terrific position to be in—the customer knows your company's capabilities and is so confident of fair pricing that it has selected you from the beginning. Or perhaps, you have a patent situation or a "secret technology" that makes your company the only one that the customer feels they can use.

In this case, the challenge is different from a competitive proposal. Your proposal must reinforce the customer's decision and deepen the trusted relationship that may already exist. If your customer thinks for a moment that you are trying to take advantage of the sole source situation, you may not get the award. So, provide the best labor rates possible, and plan to put your best people on the job. This kind of situation is worth its weight in gold—maintaining a sole source, trusting relationship is worth your very best effort and attention to your customer's needs!

Very often, the proposal is primarily an outline of how you plan on doing the work and the sign-off is mostly to be sure that there are no hidden liabilities in what is likely to be a T&M contract. Since there are no "fairness to the competition" issues to manage, your dialogue with the customer during the proposal process may be fairly uninhibited, and you and your customer are more free to work out plans and expectations that are favorable to both parties. Of course, your customer's decision to make the RFP sole source cannot be on a whim. The contracting agency's decision to proceed on a sole source basis can indeed be challenged by a competitor. The potential for this challenge should be kept in mind by both you and your customer as you work on negotiating the contract. It is just one more assurance of honest and equitable behavior on both your parts.

WINNING AND ALMOST WINNING THE CONTRACT—FINAL NEGOTIATIONS

Returning to the more common, competitive-type proposal process, we were describing earlier, assume you have been successful and got a cost competitive, technically attractive proposal to the customer. At this point, the customer may decide to engage your company and place a purchase order (PO) or award a contract—or the customer may want to clarify some of your offerings to help evaluate your proposal. Very often, in the interest of fairness, and as may be directed by the customer's internal procedures to confirm with applicable FAR requirements, there may be a dialogue with all the acceptable bidders (i.e., those that are considered capable and compliant and in the competitive range) and then an opportunity to issue a "BAFO" (best and final offer). This is a great opportunity to second guess all that you have done until now! Or to correct errors, or to reduce profit if the risks look less or the competition looks hungry. In any event, you can count on very senior management's interest in this process, and actually, you do want them to be part of it. It may be a tense period, but you are nearing the finish line!

Sometimes, the after-submittal activities are limited to two or more "finalists"—the best two offers that have been received. Since it takes time and money to evaluate and work with bidders who are clearly not in the running, less attractive proposals may be set aside while the customer works the finalists to decide which is the best offer. These meetings or correspondence are important of course—you are a finalist, and one step away from winning or losing. If concessions are necessary, they should be seriously entertained by you and your company. (Do not make them alone—make sure you have functional and/or senior management buy-in to anything "extra" that may be requested.)

It will be your job to analyze and publicize any changes from what has originally submitted. In some companies, the proposal manager may have some latitude, maybe up to 10%, but even if you have that latitude, you may want to use it in collaboration with your management—if you make 15% profit on a tough job you may be a hero—but if it is 5%, you may have wished that you did not give away two-thirds of the profit at the final negotiations!

CONTRACT REFINEMENT

Assume your bid is selected. The next step is to place the PO or award the contract—but maybe not the very next step. Oftentimes, the customer may want one more meeting (or exchange of documents) to verify that you are both clear on the scope of the contract. In theory, this is a formal process. Detailed requirements, a very specific statement of work, detailed deliverables, clear schedules, Ts and Cs, etc., are intended to completely define the work. In the proposal process, you wrote clarifications and exceptions, often on the form provided by the customer, and each one was responded to—usually in writing and signed by the customer—often by both the cognizant engineer and the appropriate procurement official. But despite your mutual best efforts to clarify the work, especially in complex procurements, there is always the chance for miscommunication. Therefore, it is not uncommon to have a "pre-award" conference. Again, this is an opportunity to build rapport and mutual respect with the customer. And there is the potential for nasty surprises—in discussion, it may be that the customer wanted more of something (work, reports, analyses) than you thought and priced. This can be a tense meeting. The customer has selected you after weighing what it thought were equal comparisons of scope with your competitor. If there is a large disconnect, the entire program may need to be re-bid—which would be unfortunate for everyone—except maybe for your competitor! There may be a need to make concessions to close the deal. You will need to be very careful at this time. Your desire to get the contract is so strong you can taste it! By no means should you accept significant costs or risks at this time, no matter how tempting. You will also want to involve your senior management, so they do not think later on that you gave away the store in negotiations!

This, too, is an opportunity to work with the customer. If in one area the customer wants a little more than you proposed, there is a chance that in another area, the customer might want a little less. This is the essence of negotiation—finding areas that do not hurt you too much to give up, and, equally important, finding areas that your customer can give up without too much difficulty. Successful negotiations are ones in which both parties are satisfied. Unfairly leveraging vulnerability against you or you against the customer will only breed resentment and greatly reduce the chance that you can work together through future challenges on the program.

At the conclusion of these negotiations, finally, you may get that PO or contract award that you have been working so hard to secure! Be sure you read it thoroughly—and have your contracts department do so also. While it is very unlikely that your customer may try to "slip something in," it is more likely that an error may have been made. When you do get the award, you will have to review and accept it—this is a key point, because it is in this acceptance that you agree to what's in the PO or contract award <u>only</u>. Informal agreements are usually specifically abrogated by the PO text. You will have to sign, most likely, that there are no other agreements or understandings that are not covered by the PO. Anything that was not clear and was resolved in negotiations needs to be written and formally included in the PO or its attachments. Even if you "trust" your technical counterpart's word, the customer may not have the ability to defend the informal "clarification," or the customer's representative may be reassigned to another project, and in fact (and in law) all that binds the party is the written PO.

BUT WHAT IF YOU LOSE?

We have spent some time talking about the endgame and reaching and documenting clear agreement. But, of course, you should not expect to win every program you bid on. This is never a happy event—work that you thought you would have, that you arranged staffing for, that you planned on being an important part of your company's (and your) future, just is not going to happen! What then?

The most important thing at this time is to find out why your bid was not selected. The FAR calls for "losing bidder's information," and as a minimum you can expect to find out how many bids were solicited, how many were submitted, and who the successful bidder was. You will also find out the contract price.

But in addition to the formal disclosure of these details, very often customers (especially those with whom you have good working relationships and expectation of future work) will invite you to review the reasons you were not selected. Most often, they will disclose their evaluation process, perhaps how many "points" you got for each category evaluated—categories such as technical content, program office, and price. This outside look at your proposal is very useful. It is a direct insight into how your company compares with its competitor (at least in this one instance). If you were the proposal manager (as we have been assuming), your detailed knowledge of the proposal is very valuable in understanding the loss. In order to profit most from the learning of this conference, you may be asked (or you may volunteer) to write a "lessons learned" paper or to do formal or informal briefings of your peers and management. After all, your team put a lot of effort into the proposal, and you want to recover as much good from losing as you can.

And, as a leadership note, don't forget to thank your team for their work. They will, of course, be very interested in why the team lost, and you owe it to them to brief them promptly and directly as well. As disheartened as you may feel, some of them may be as affected as you, and the only solace you can capture for the team is found in a frank review of what you did right and what you did wrong in this proposal.

As your colleges (or you yourself) go forward on new similar proposals, especially with this same customer, take some time to review past unsuccessful offerings. Share your knowledge and learning with the next proposal manager. This help may make the difference between winning and losing the next opportunity. Turn this loss into a tactical loss, in the hope of making a strategic win.

WHAT IF YOU LOST FOR THE "WRONG REASON?"

As you learn more about the bid process and about who won, you may feel that the bid was in some way flawed or unfair. Examples of this might include the following:

- Your competitor had inappropriate "inside information."
- Your competitor purposely underbid, expecting to "make it up" on changes.
- Your customer made an error in the evaluation of the bids.
- There was some misrepresentation in your competitor's offering.
- Or, there was some other anomaly in the bid process.

If you feel you have adequate grounds for a formal protest, you should carefully consider this action. As proposal manager, you are one of the most likely people in your company to know or suspect an impropriety. However, the decision to enter a formal protest is most likely to be made by senior managers in consultation with your legal department. Protests can be messy, and as you can imagine can alienate your customer/potential customer. However, do not fail to alert your management to any appearance of bid impropriety. If, indeed, a competitor has been awarded a contract based on an impropriety, it may be possible to set things right and possibly cause a rebid. If rebid on an even playing field, it may be possible to snatch victory from the jaws of (apparent) defeat.

END OF CHAPTER QUESTIONS

For Discussion

1. Your potential customer has announced a bidders' conference for a current, complex RFQ. As you prepare for the meeting, what should you consider as your goals and cautions?
2. Suppose it is your job to establish a win strategy. How would you start? How collaborative could/should it be?
3. As the proposal manager, your marketing department has developed a PTW. Great! Now all you have to do is get your price to that level, right? Really, is it that simple?
4. Discuss the formation of the proposal's vision and theme. (Remember, you may own the task to create the vision, but you want the team to own the vision!)
5. As the price is being developed, some functions seem to be too aggressive, and others appear to be too conservative. How would you challenge them and/or evaluate your own thinking relative to their estimates of the scope?
6. You are about to present your proposal to senior management for sign-off. Your company has an expected template for this review, but your job is to communicate why the proposal is worthy to be submitted and why the pricing is correct. Beyond the template, what can you do to ensure a successful sign-off?
7. Assume you have submitted your proposal and have made the final cut—it is your company against two others. You have been asked to develop a BAFO. What thinking should go into this process?
8. Hold a simulated "post-award" conference with this premise: The customer is concerned that the bidder is looking to find loopholes in the specifications, and the bidder is concerned that the customer's expectations go beyond what's specified.

Written Assignments

1. Describe some effective ways of gathering information to influence a bid–no bid decision. Give some examples of legal approaches to getting information related to the proposal and your competitors' strategy. What steps or practices would be unethical?

2. Suppose your potential customer has not identified selection criteria. What aspects of your proposal are likely to be most important to him or her? Is there some appropriate way that you can explore or infer what these might be?

3. Your proposal team is being formed, but you have had a limited say in who is assigned. What considerations would you use in dividing up the work?

4. Create a "risk register" for a sample program of your choice. In a real proposal, what steps or strategies would you use to identify and then evaluate these risks?

5. Describe each of the reviews that your proposal will have (Green Team, Red Team, Black Hat). What should be the outcome of each review, and how would you manage the introduction of that outcome to your team? What if you and/or your team disagree with the reviewers' comments?

6. Assume you have been selected as a result of a competition for a complex program, and the customer wants to have a post-award meeting. What are your goals at such a meeting?

7. If your proposal is rejected in favor of a competitor's, what steps should you take to learn about why you lost? How do you deal with communicating the criticisms, some of which are very direct, to your team? To your senior management?

8. If you lose a bid, under what circumstances should your company issue a formal protest? What would be some of the valid, and some of the invalid, reasons for such an action? What would some of the ramifications on long-term working relations be if this is an ongoing customer on other programs?

Mini Project for a Team

1. Develop the storyboard outline for a design program for a project aligned to your company's normal business. Create the boards and present them to the class.

6

Planning the Program and Starting Work

Now that you have won the contract—maybe there was even a party to celebrate the win—it is time to do the real work. The work that you did in preparing the proposal will be the starting point for the program. Schedules, plans, estimates, risks, and opportunities are all part of the proposal record. Ideally, and especially if there was little pre- or post-selection negotiation, those plans will form a good basis for the way that you will organize and run the program. Some companies require very detailed and highly reviewed plans as part of the proposal process, which if you win, are invaluable in getting started. However, if you lose the bid, they are essentially wasted effort. In any case, you will not be starting from ground zero as you kick off the program; the proposal, with whatever detail in which it was prepared, is the basis of your execution plan. It represents your interpretation of the customer's requirements. In most cases, the proposal is noncontractual—the purchase order (PO) or contract and its attachments rule the program. But as a minimum, your proposal is a good-faith record of what you offered and expect to provide. Sometimes, you can have it included as a reference in the PO, and this can be helpful if misunderstandings arise. So with a firm understanding of the PO and its attachments, and with your offer vision (in the proposal) in mind, you are ready to begin the program.

You will find that the start-up effort really divides into two categories:

1. The management part
2. The leadership part

As we discussed earlier, conservative, traditional companies tend to be management (versus leadership) oriented. But, successful program managers (PMs) are both managers and leaders. And since your probable corporate culture is more management-leaning, you probably need to drive yourself to be more attentive to leadership issues. The nuts and bolts of starting up the program are generally documented in formal procedures. As a PM you should use your procedures as tools—they will be very helpful in the mechanics of getting the program started. Figure 6.1 shows a PM comfortable and relying on his procedures. Reliance on your procedures is a best practice. However, overreliance on them at the expense of motivating your team is not. Do not follow detailed procedures at the expense of energizing, helping and inspiring your team. Figure 6.2 shows a PM setting a performance goal with a team member.

Remember that a successful PM uses both his or her left and right brain. Most of your colleagues will be left-brain oriented. You can just look at your procedures to see this is the dominant culture. And of course, there is great tendency in all of us to align with "what is expected." This means you should have no problem getting into the left-brain work—scheduling, allocating, planning—after all, the products of this thinking are usually mandated. You probably will find not find any procedures on team-building and motivating a nascent team. So if part of the effort is well

FIGURE 6.1
A PM comfortable with his procedures.

(From Dollar Photo Club, File #49968896 texelart.)

FIGURE 6.2
A PM setting goals with a team member.

(From Dollar Photo Club, File #62632519 texelart.)

mapped, and part is less understood, should not you concentrate on the "softer" issues—even though (and maybe especially because) they are not as simple?

With this caution, let us consider what you will be doing in those first few, terribly important days of the program. Remember, as with virtually all activities in life, getting started is the hard part! If you have had other obligations while the customer was evaluating proposals, it may be tempting to finish those up—they will likely have clear tasks that have to be done, and there is a comfort and satisfaction in doing those "clear" tasks. You would probably like to bring that work to a nice, neat conclusion. But resist the urge—and get to work on your new program. Find someone to whom to delegate those interim tasks. And get started! At the beginning of a multiyear program, it may seem like a day or two starting slow is not a problem. But it is—as you approach each deliverable, you most certainly would like to have that day or two back. And there is a more insidious and more severe problem—these first few days will be setting the tempo for the program, both in the minds of your managers and your team members and also in your own mind as well. Act with urgency from the very beginning and you will be setting an example for the entire program.

THE MANAGEMENT PART

Before you or anyone can begin formal work on the program, you must establish a charging mechanism, and budgets or accounts. If it is a large program, this is nontrivial work. In the proposal process, each department (system engineering, design engineering, manufacturing, etc.) will have provided a detailed estimate of the work that it is expected to do on the contract. You and your financial analyst (or program analyst) will write formal work statement authorizations that carve out program funds to authorize that work. These will be organized in accordance with the work breakdown structure (WBS) that is likely to have been done as part of the proposal process. The WBS is an organized indented list of the work that has to be done. It can be (and usually is) the basis for the detailed task schedule. One important lesson to learn when creating structures and schedules is to make the ownership of the tasks clear. No one on the team will give your schedule the attention that you would hope—but everyone will look for his or her name. People want clear direction. They want to know "What do I have to do and when must I finish it?"

It is common to hold back a portion of what the departments quoted as "management reserve" (MR)—with the idea that if they work to 90% of their estimate, and they have a problem, you have 10% in reserve to help them. This is essentially a reasonable approach, but it does cause some problems:

- Since it's so common that 10% will be held back, some "smart" estimators will just pad their quote by 10%, so when they get their allocation, they will have an easier time living within it. Well, the risk is obvious—your proposal could be 10% higher than it should be, and you could lose the job for that 10%. The remedy is to work with the estimating organizations during the proposal phase to be sure their quotes are adequately aggressive.

- When you do allocate the funds at 90% of the quoted values, you are likely to get some pushback—the quoting manager will tell you that the estimate was not padded and wants the full allocation. This is another sticky issue—and it can be exacerbated by company culture. If people are reprimanded for overruns, then the best way to make sure that one does not overrun is to quote high!
- Although it is frequently undervalued, it is the case that this estimate is, in absolute fact, <u>an estimate</u>. It is very likely that the margin of error in the estimate is over 10%. You and the estimator both know this, but "the game" requires you to pretend it is ±0.0%.

This is a great example of where management needs to take a back seat to leadership.

The theory is that a contract for work is forged between the PM, who gives the funds, and the operating (or functional) manager (FM) who is responsible for the work. It is based on a clear understanding of what has to be done. If the FM does the work for less than the agreed number of dollars (or hours), he or she is successful. If he or she exceeds the allocation in doing the work, he or she has failed. So, why would anyone not want to make the allocation as large as possible so as to be "successful"? Only the need to win the job would temper this opportunity to "feather one's nest" (that is, make the goals overly comfortable). And if there are an abundance of other opportunities, and the FM is not in need of the work, he or she can try to get all the feathers they can! Furthermore, the probability that the work is fully understood at the time of the quote is essentially zero. There will be both positive and negative surprises. If the FM gets a positive surprise, he or she may just smile and say nothing. But if a negative surprise is encountered, he or she is likely to say "I never quoted that!" with the implication that it is your fault. The next step would then be to ask for some MR or, worse, expect release from the "contract" to do the work for the hours quoted. There is no "management" solution to these problems, because, in theory, everything is known at quote time and there are no surprises!

At times like this, true leadership is necessary. As PM, you and the FM have to feel like "you are in this together" and the only success is doing the work and meeting the requirements for as few hours as possible. Dozens of jobs have been successful because of these relationships, and several have failed because of bickering over allocations. Even in jobs that were not successful, the usual argument is "they overran their allocation" countered with "the scope was not clear (or changed)." It is virtually never the case that it is one or the other—so the possibility of a clean indictment of a nonperformer is very low. All the bickering, therefore, is a waste, and a distraction from doing the job. Establishing the "we are in this together" environment on your program is one of the most important things you can do to ensure its success. (In fact, with programs that run into serious problems, <u>all</u> the managers are adversely affected—it is never blamed on one individual—except gasp—sometimes the PM. And if the PM tries to use the defense "he or she overran the allocation," all you hear is that the scope changed—and since it always does—you lose!)

So, if the program is set up as "you better get this done for 0.9 × hours, or else," you are in deep trouble. Leadership based on threats is doomed, and in fact, it has no business being called "leadership."

Also part of the basic program plan is the establishment of a "risk register" or a "risk and opportunity register." Here the foreseeable financial risks to the nominal execution of your program are tabulated and quantized. Since these are possible events, the probability of their occurring is also tabulated. Thus, the size of the risk is quantified. The same is done for "opportunities"—things that if they go better than planned could result in a positive impact on your spending. Also, the anticipated date that the risk or opportunity will either be realized or avoided is entered. During the proposal phase, these risks and opportunities are valuable to guide pricing—risky programs might have contingency money added, or higher profits quoted (for fixed-price bids). Thus, the open risks are always in front of you, to manage away (risks) or to make happen (opportunities).

Table 6.1 is an example risk and opportunity register. The net dollars should be covered, if not exactly equal to your MR.

And thus, the management part of the program start-up is generally intricately choreographed by your company's procedures. What the artifacts are, where they are stored, who has to sign off on which ones, how they are maintained, how schedules are baselined and monitored—these are all covered in your procedures. But checking the box for all your management artifacts, while necessary, is certainly not sufficient for your program's success. That's where your people skills are needed.

TABLE 6.1

Risk and Opportunity Register

Risk or Opportunity Number	Risk or Opportunity	Value	Probability	Net Value	Date for Resolution
R1	If the board designs are late by one month, the schedule will be extended and the level of effort (LOE) costs will rise.	$45K	0.2	$9K	May 1, 2016
R2	If the shock test fails, the structure will have to be redesigned.	$200K	0.1	$20K	Nov. 15, 2016
R3	If our costing rates go up by 10% in 2017, our costs will increase.	$100K	0.1	$10K	Jan. 1, 2017
O1	If we finish testing one month early, our LOE will decrease.	–$45K	0.1	–$4.5K	May 23, 2017
O2	If our costing rates go down by 10% in 2017, our costs will decrease.	–$100K	0.05	–$5K	Jan. 1, 2017
Totals		$200K	—	$29.5K	—

THE LEADERSHIP PART

For this discussion, let us assume that the program is a design and development program, wherein you will design an electronic system (i.e., radar, sonar, instrumentation) and then test and qualify it—that is, verify that it meets all of the design and contractual requirements. It is likely that you will be able to have some of the people who worked with you on the proposal assigned to the implementation team. This is obviously a big help, because you formulated the successful proposal plans with them, and thus you probably have a shared vision of how the program should be run. This "vision" includes reporting relationships and schedules, but it also includes the softer issues, such as how proscriptive the technical direction should be, what kind of reviews are planned, who will do what part of the leadership function (you or the leads—electrical, software, etc.). Often you as the PM will work with your "dotted line staff," the functional leads, to resolve these issues. They will be the "sergeants" of the program—working with the designers to be sure the progress is on track, and that the technical issues are known and are being addressed. As with any good team, it is important to find complementary talents. For example, the weaker the PM is technically, the more important it is to find leads who are technically superior. Again, this is likely to be an extension of the proposal team—the software engineer who envisioned the system and/or wrote the proposal text for the software, is likely to be the best candidate for the software lead function.

However, it is very important to realize that the talents of creativity and leadership may not necessarily be found in all of your proposal colleagues. While the mechanical engineer on the proposal may have developed an excellent technical approach to the system's requirements, this engineer may not be a leader "by nature." If there is a team of five mechanical engineers who need to implement the vision, it is important that their leader can communicate with them, solve problems, and detect and resolve problems as they evolve. Just because an engineer has had a great and maybe contract-winning concept, it does not mean that this engineer knows how to get the most out of the sub-team. Again, the principle of allowing people to work to their talents should protect you—in general, people without leadership talents, given the opportunity, will avoid leadership roles. There is one cultural issue, however, that can upset this tendency: if leaders are more respected (and better compensated) than great technical contributors, there is a risk that some ambitious engineers may feel the path to career advancement requires leadership experience. Fortunately, many companies recognize this difference in their employees' capabilities, and provide a "dual path" career progression. If well executed, superior technical contributors can earn as much as superior managers. So, hopefully, you work for one of those enlightened organizations, and your leads really want to do leading, because their talents align with that role.

Absolute trust and respect between you and your leads are vital to success. If you suspect a problem (for example, mathematical modeling purity seems to be taking precedence over schedule demands), you must be able to ask your systems engineering lead to look into the concern and either determine whether everything is fine or take steps to make it fine. In a large program, you cannot see and correct all the problems, and you cannot review and adjudicate all the actions of your leads.

I do not travel much but much can be learned by listening to people from other cultures discussing management issues from their viewpoint. On a plane from Barcelona to Frankfurt, I sat next to a Spanish businessman. The European Union was in economic distress, and there was a good deal of blame on the Continent. The Germans were doing well economically, and felt that their work ethic, when compared to that of their Spanish friends, was the reason. Maybe they thought the different workday hours kept in Spain (with a break in the middle of the day, offset by a later quitting time) was the reason. Whatever the cause, this gentleman felt that the Spanish had unfairly developed a reputation of their culture having a built-in bit of laziness. In fact, this gentleman assured me it was the opposite. He felt that the typical Spaniard worked more hours than the typical German. Then what was the reason for the greater economic success of the Germans over the Spanish? The Spaniard sitting next to me had a different understanding of the cause. He said that in Spain there was a general distrust of the work being done in other departments, and much time was spent reviewing and or second-guessing the work of colleagues. He thought that in Germany, where "organization" is king, the expectation was that everyone knew his or her job and did it, and no one had to check. While this is obviously a gross generalization by a single individual (whose only known credentials were that he had a seat next to me on a plane), it does point up the large overhead penalty paid in non-trusting cultures. If you require your staff to justify all their actions, you will spend a great deal of your time, and theirs, in this process. And it leads to reduced initiative, since actions may have to be justified, while inaction is harder to notice.

So as you form your team, typically starting with the leads, it is important to build the vision at the same time. As mentioned before, you are unlikely to have complete control over the people assigned to your program, but you may have more than you first think. There is nothing like adamancy and persistence in getting your way. (Remember, getting "your way" on staffing issues is not an ego event; rather, you should see it as your obligation to your team. Building a strong team is a benefit to the strong team itself.)

As you add your leads and as they work with you in adding the engineers to your program, it is important to communicate and refine the vision of the program. A "vision statement" can help communicate this vision to the forming-up team, and it helps you clarify your own thinking.

But remember that the vision is a living entity and should evolve and grow. And remember that you do not own it by yourself. Indeed, a joint ownership of the vision by all members of the team should be your goal. The more collaborative you are in establishing the vision, the more your team will buy into it and succeed. Remember the quotation on leadership from Lao Tzu mentioned in Chapter 1:

> The wicked leader is the leader who the people despise.
> The good leader is the leader who the people revere.
> The great leader is the leader of whom the people say, "We did it ourselves."

The more you make the vision the team's vision (not your vision) the more successful you all will be.

You definitely want a team that can work well together—as you add people to a growing program the best way to add people is to ask the people currently on the program about the candidates that you are planning to add. Listen carefully to their response—you may want both Engineer A and Engineer B on your program, but if Engineer A and Engineer B do not trust one another, you may have to make a choice. In so doing, make sure you consult the rest of the team as much as possible.

Does all this checking and talking take time? Of course it does. But it is probably the best place to invest your time at the start of the program. You may have to decide if you should spend a half hour talking to your team or a half hour filling out a report—the report may not even be read—spend five minutes on the report and concentrate on the people!

I had the good fortune to start up two similar programs, one six years after the first. I thus had the opportunity to apply what I learned in "vision management" from the first program to the second.

First program (Program 6): At the time we won this program, there was an unhealthy adversarial relationship between engineering and operations. It was characterized by a general lack of respect for the quality of work done in each department. So I had an idea—set up a part of the building where everyone working on Program 6 would sit together—engineers, draftsmen, factory planners, wire harness developers, board fabricators, software engineers, etc. The principle I was trying to apply was that people working together in close proximity toward a common goal will tend to form a true team; and the physical separation of the different departments leads to a lack of respect for the work and work ethic of the other departments. Co-location was pretty much a brand new concept for our company, and it was (and still is) a great idea. So I set about implementing it—I sold it up the line to the plant manager and to the FMs involved in staffing the project. Then I told the team, "Hey, great news, we are all going to be sitting together in one big happy program area! Isn't that great!" Well, not everyone thought it was great. I heard complaints from members of "my team":

- "Now I have to walk 50 yards further to the coffee pot!"
- "I have sat next to Jim for five years, and I like talking to him."
- "But there are no carpets in the new area!"
- "Now I have to walk further to lunch."
- "It will be noisier there! I need quiet to think."
- "But there are exposed pipes on the wall!"

I could not believe it. These all seemed to me to be the most trivial complaints—I mean, we were going to get to sit together! I had battled the expected seating arrangement bureaucracy and won!

In the soul searching I did after the cold reception to the idea, I learned why. It was <u>my</u> vision, not <u>our</u> vision. I had successfully sold the idea to the plant manager, but I did not even discuss it with the team! We did all move in together, and it did all work out well eventually, but it could have been better from the start. So I learned....

Second program (Program 9): Program 6 had entered a stable production phase and we won a new program, Program 9. As with Program 6, there was a need to work closely among the different disciplines in the design and development phase, and I thought it would be good to co-locate as we did on Program 6. But this time, the operative word was "we" and not "I." As team members were added, I took time to discuss the idea of co-location with them, and they discussed it among themselves as well. While I thought it was important to have absolutely everyone in the same area, the wisdom of the team prevailed, and we let the software engineers sit off by themselves in a closed area. And when it came time for me to move to the co-located area, I was shocked to hear that it would not be necessary—I should stay up in the front of the building working with senior management, while the "real workers" moved to the common area. Well, this was an ego-blow—how could they manage without <u>me</u> there! I listened to the team and left my office where it was—and the team was co-located and did spectacular work. We had on-time deliveries, full profit, and a happy customer. Ironically, at the time I hadn't heard Lao Tzu's leadership vision—*The great leader is the leader of whom the people say, "We did it ourselves."* It is ego-threatening to "disappear" into the shadows, and if I were a "great leader,"

I certainly did not feel it at the time—it did not really feel like I expected great leadership to feel... but putting the team first, ahead of your own self-image and ego, is in fact the essence of great leadership.

SOURCING

Engineering Labor

At the beginning of a program, especially a large program, you may find it difficult to get all the people you need assigned as quickly as you need them. This is even more important if you are going to be "selective" as we discussed above. One solution is to borrow people from other divisions; another is to use contract engineers; another is to wait until the right people free up from their current assignments; another is to ask for overtime from the people you do have; and still another is to outsource complete work packages. If you think of the work as just the number of hours that are required to do the task and do not pay attention to the consequences of these approaches, you can have a poor outcome.

People Brought in from Other Divisions

This approach may be useful. The people brought in may not have your local organization's success at heart, but you can count on some of that hunting party affinity to integrate these folks with your core team. At least, they are employees of the company with a career stake in the program's success. Be careful, however of costing rates. Depending on your company's disclosure statement (the accounting basis disclosed to the government as to how you treat overhead and other costs in your rates), you may pay a premium for their services. Sometimes you must pay for their hours at their regular rate plus their G&A (general and administrative) markup (G&A costs are adders to labor costs that cover such things as office space, utilities, computer services, etc.—in some accounting systems, other divisions of a company, when they do work for your division, may have to add G&A costs to be consistent with their disclosure statements to the government.) This can make for an unanticipated cost in your program's execution. So those folks need to be worth the extra premium you may pay for their hours—which could be on the order of 20%!

Contract Engineers

This is a common solution to insufficient staff at the start of a program. It comes with a lot of pluses and minuses, and they are worth noting. Firstly, no "contractor" is going to own the program the way your core staff owns it. They are in it for one thing only: money. While your core staff is there for the money also, they have additional motivations that will help the program—building their reputation, earning a promotion, getting a raise, helping their colleagues, etc. Even the best contractor does not have these motivations. As contractors, without benefits, they may get a higher (or even much higher) hourly wage than your core people. If (or probably when) your team finds out how much they are making for essentially the same work, you will face a potential morale problem. The disparity in salary may make it difficult for some of your engineers to work with the contract engineers: "If you are getting paid so much, why do I have to help you figure that out?" may be the unspoken interface. These are major drawbacks, but there are some offsetting benefits—although contract engineers may get a high hourly rate, they may cost your program quite a bit less than your company's direct employees. This is because certain overhead costs (management salaries, vacation, benefits, office space, etc.) are often not ascribed to their contracts. Often they are treated as material—and the material burden (the percentage added to material your company buys to execute the program) may be significantly less than the overhead rate. Another benefit is that they serve "at your will." (It is not an exclusive "your"—other folks, including FMs, may likely have an input.) But this means that when you do get your permanent staff, they can be released promptly. However, while that sounds good, when they leave they will take what they learned about the program with them, and you will have to pay for a new person to come up to speed. And some very talented (and/or very clever) contractors can work themselves into positions on your program that make it very painful, if not practically impossible, to terminate them. Further, your company may have rules that govern how long a contract engineer may stay—so you might develop a dependency on a contractor and have to lose him or her anyway. Now this does not sound that good—because it is not a good situation! You may have to be a little selfish at this point—let some other program use contractors while you get the permanent employees! But remember that when you are "selfish" for the program, you are doing your duty to the program and to the people on your team.

Waiting for the Right Engineers

This approach is attractive in the long run, because you get the right people on your program. But it has pitfalls too. First, while you are waiting for a particular engineer, the work is not getting done. You fall behind on that task, and furthermore, that task might impact other tasks. And often you may hear: "Chris will be free in two weeks." And you (foolishly) believe it. If you just wait two weeks and look for Chris then—you may be very surprised. As it turns out, Chris's task was underquoted—or the circuit blew up, or Chris was sick, or the customer made a change, or any number of potential delays. You may think you have a commitment for Chris to start at the end of the two weeks, and the FM may have been very sure that Chris would be available, but, just like in court—"possession is nine points of the law." That is to say that the program who has an engineer is likely to be able to hold that engineer until the task is done, regardless of what was expected or what the calendar says. Generally, the program that has Chris will benefit if it goes to senior-level management arbitration, which by the way, is not recommended. Better to work out something with Mary, her manager, and your team to tolerate the delay. And better still to have a backup plan to implement if that two weeks turn into two months. Waiting is dangerous!

Overtime (OT)

Most companies today pay their salaried employees overtime (OT) at the standard hourly rate. Sometimes, a certain amount of "casual OT" is expected from engineers, especially if it is of short duration or from more senior and thus higher compensated employees. But remember, it could be a violation of your disclosure statement to force or expect casual OT, and it may be unfair to the employee. True, some people really enjoy their work and are driven to success, and it would be nice to think that this will allow them to "get ahead"—and it might. But the emphasis is on the word *might*. Engineering work, especially highly creative engineering work, is hard to measure. So an employee putting in 20% OT might not produce an obvious 20% more work. Pat may be reluctant to self-promote and tell folks she has been working a lot of OT. This reliance on casual OT is thus fraught with peril, including the peril of disengaging some of your highest potential engineers. Remember, one of your most important jobs as PM is to make sure the people are engaged and enthusiastic. The best way to kill those feelings is by unfair treatment.

But What about Compensated OT?

There are a lot of pluses with fairly compensated OT. Compared to bringing on another person and suffering through (and paying for) his or her program-learning curve, the folks already on the program are free of that burden. Out-of-hours work, in the evening or weekends, can be more productive, because many of the distractions of the regular work day—idle chatter, administrative memos, noisy nearby discussions—are much reduced in out-of-hours work time. But out-of-hours work has its drawbacks also—the people that might be needed for consultation or information might not be around, and some things will thus have to wait till the next day, or may have to wait till Monday morning. But the compensation angle is a big plus: some of your folks may be buying a house or planning an extensive and expensive vacation, and would be very happy to work OT. Compensated OT is way less unpopular with your engineers' spouses—it is always important that the engineer be appreciated at home, also! But remember, there is a limit as to how much extra you can get and how long you could expect to get it. People really do have increased stress from extensive OT, and if you overstress your team at the beginning of the program, you may not have the ability to put in the extra effort when you need it as deadlines approach. I would recommend no more than 8–10 hours per week and even then for no longer than a couple of months. There are always exceptions as to what people can or should do, but remember the stress factor. And always avoid the "mandatory OT" crutch. While it does send a message to the team that extra effort is needed, the issues involved—how much, when does it need to be done, possible resentment, possible inefficiencies, and possible disengagement—make it almost always a losing proposition. If you make it "mandatory," and even one person cannot or will not do it, you will have a problem that does not have a solution. If you cannot enforce a rule, the respect capital you lose is a significant problem. It is way better to talk to each of the people separately and work out a plan for extra effort with each of them, individually.

Today's programs are not terribly socially different from hunting parties of millennia ago. Humans are at the top of the food chain largely because they can communicate and work together, and humans knew how to plan then, too. Figure 6.3 shows some early planning documents for a hunting program in 20,000 BC. So if you are a chieftain organizing a hunting party in the year 20,000 BC, you realize how important it is to select the right people for the hunt. Your very existence and that of your family and friends (your tribe) depend on the success of the hunt. Avoidance of starvation is a powerful motivator—maybe a little bit more so than is an opportunity for promotion or for a salary increase. The instinctive social interactions that form our anthropological roots are powerful forces to consider in team selection and team building. "We are all in this together" is as true now as it was thousands of years ago (though the consequences may be a little more benign—you probably won't starve if your program fails—but it will hurt nonetheless!)

If your hunting party is large, it is unlikely that you will know all the tribesmen and tribeswomen and their talents for the hunt. You will want someone who can track, someone who can throw a spear, someone who can find the way home, etc. Your lieutenants (your leads) will help you find the right folks with the right skills. You will need to rely on them. The best lieutenants will understand team building and will make sure the people selected are competent—*and that they can work together!* Mutual trust and respect among the hunting party are vital.

FIGURE 6.3
Early program-planning documents.

(From Dollar Photo Club, File #33565226 kaetana.)

OUTSOURCING WORK PACKAGES

One technique for getting more work done sooner is to take tasks or subsystems and give responsibly, under subcontract, to an outside organization. As the other "solutions" mentioned earlier, this one has its own benefits and issues. One obvious issue is that it takes time to specify the work that is needed, to select a supplier, to flow down all the necessary requirements of the contract, and to monitor progress. If your company already has a "go-to" organization, one that has proven itself trustworthy and competent, many of these issues are minimized. But even under the best circumstances, the knowledge of the product that is gained by their engineers is going to be lost to your program when their work is done. As their work is integrated with the rest of the program, interface issues may arise, and you will either have to bring them back (involving another round of contractual administrative distraction) or have one of your own engineers learn enough about their design to solve the problem. Either of these is expensive and possibly will delay the schedule. Personally, I find this one of the least attractive solutions to early staffing, but as with all the other solutions, it may have its place under certain circumstances.

IN SUMMARY

There is no simple answer to dealing with understaffing at the start of the program. And that is absolutely true. A blend of these strategies may be the best way to move forward as expeditiously as possible. But similar to every problem your program faces, the more you involve the affected people in the solution, the better the solution will be, and the better the people will feel about the solution. Being part of the decision is the best way to make an individual be a part of the team. If you ever question the value of including your team in important decisions, personalize it—how do you like it when a plan that directly involves you is pushed upon you with no opportunity to contribute to it or improve the plan?

OUTSOURCING PRODUCT

We talked about sourcing labor above, but we should also discuss issues with the outsourcing of product.

Depending on the capabilities of your company, you will likely need to make decisions as to whether to have the product or parts of the product built in-house or by a subcontractor. This can be a critical and tricky decision. This "make-buy" process may be formalized in your company, and, regardless, there are a new group of interested parties in the decision, all with varying degrees of formal and informal power. Your job is to understand the benefits to your program of one approach or the other and lead the interested parties to that solution. In theory, this sounds simple. But in fact, it almost never is. A manufacturing group may want the work on your program to increase its hours and keep its rates down. If they are too desperate for work, you may find that your program has attracted a great number of helpers and watchers. But on the other hand, it is extremely efficient to be able to have your engineers walk out on the shop floor to see and resolve a problem promptly. If a problem arises at a subcontractor, the logistics of getting the right people to the problem to evaluate it, and the time it takes to provide contractual direction to them to make the change, can be very damaging to your program. Plus, some subcontractors may take this opportunity to prey on you, charging unreasonable change costs, which you must detect and disallow. This is likely then to cause friction between your organization and the subcontractor you expected (and needed) to be a trusted partner in the program.

Every company is different, and every company's situation changes over time. Your decision (or the part that you can influence) is also based on the shop load (meaning availability) of resources in your own company and on the availability of trusted subcontractors. Remember, too, that you may not be able to choose the subcontractor. Very often, the paradigm is that the program's engineers provide all the information that is needed to manufacture a quality product in drawings and specifications, that package goes out for bids to "qualified" subcontractors, and a winner is chosen, largely based on price. Thus, it may not be possible to avoid subcontractors who have a tendency to "milk" (take advantage) the changes or ones that are on the other side of the country.

So how do you protect your program from these issues? Again, this is more a leadership than a management issue. Your job—to get the best supplier, internal or external—for your project is limited by the "power" of your office. Your "power" is limited by procedures and rules. But your *influence* is not limited. By using relationships that you have built with your involved colleagues and by listening to their points of view, you can align the forces necessary to get the right decision. Relationships are the key. Relationships with other employees who, like you, want to do what is best for the company. Of course, everyone has their own best interests in mind, but the best employees (and your company is populated with great people—large companies can be selective in whom they hire) will have a natural tendency to work with you for the good of the program (or the good of the company).

So this is great if the involved people are your "work friends," but every day you must work with colleagues, some of whom you never even met. Relationships matter there also—approaching someone sincerely and openly can make an instant ally. Your people skills, quickly identifying and working with those allies, are the most important parts of your PM skill set. The best leaders can make allies out of almost everyone—the worst leaders make enemies wherever they go.

Assume then, that through your leadership and management skills, you have a subcontractor ready to work on your program, producing, for example, printed circuit boards of various styles. You must have a way to monitor the vendor's progress and the quality of the product. Again, in large organizations, much of this work has procedures to follow or is considered at least routine. A subcontract manager and maybe one or more operations engineers will be assigned to your program—you may have a part-time quality engineer who is familiar with the product and or the subcontractor. They will all likely know more about the details of your product, more about the subcontractor and its quality system and more about the detailed schedule than you do. Your job here is to make sure the best people possible are the ones assigned to your project. Is this selfish? Yes! It is your job to be selfish for your program.

But even if you have great folks on your program, you still need to be watchful and helpful. Because of the tendency we all have to avoid areas in which we are not comfortable, it is a best practice for you to know your people and what things they might avoid. The best leaders will coach team members to pay extra attention in their discomfort zone, the adequate leader will compensate for their weak areas, and the poor leader will not notice the problem. When the people are assigned, sit down with them and discuss how the subcontractor's work will be managed.

One thing you may notice is that the discussion will likely involve statements such as: "We will get weekly reports from the subcontractor and evaluate them." This is the time to be non-threateningly inquisitive: "So, Mary (the operations engineer), is that your job to evaluate the report or is it John's (the subcontract manager)?" You, John, and Mary may all think you know the answer to this type of question, but you may all have different answers! Sometimes this type of question is not asked because it may seem "stupid"—everyone should know that! But surprisingly, those "stupid" or "naïve" questions can clarify things greatly and lead to surprising insight. It takes some humility and/or self-confidence to ask these questions, but humility and self-confidence are key leadership ingredients. Cultivate and use them!

I learned a lesson a long time ago from a customer on a complex technical project. Regular progress meetings were typically performed by a two-person team: a buyer and an engineer. But in this one case, the buyer came alone. I thought this was not the right idea. As it turned out, however, rather than a do-nothing meeting, she accomplished a great deal. She taught me "The Power of the Stupid Question." (I have since refined the concept to the less pejorative: "The Power of the Naïve Question.") As the buyer did not have a technical background, I expected little scrutiny on our technical progress. Ironically, the opposite was true. Because she was not uncomfortable asking quasi-technical questions, ones that might have been considered too obvious by the engineer, she was able to gather a tremendous amount of information about our progress and problems. Engineers (myself included) love to explain things to people that are interested. By being confident enough to ask a wide range of questions, and playing upon the desire of the technical people to explain what they were working on, it was one of the most effective fact-finding missions that I ever witnessed. (Fortunately, the program was doing well at the time of this checkup—otherwise, perhaps, I would not have noticed or learned the "Power of the Naïve Question.")

BUILDING THE PROGRAM CULTURE

Start-up is the time to build the program culture—that little bit of the corporate culture that you and your team can carve out and make your own. As described earlier, this culture is best established collectively. As your team forms, get together and talk about what wonderful things you will be doing together. Ask people to talk about their professional interests, and how the program can help them achieve some of their work goals. Surely some of these will be private, but when people talk about what they would like to do and not do on the program, you can ask questions such as—"Is that something you have always wanted to do?" "Has anyone else here done that—or maybe you know someone that has? How did that work out?" The more you expose your own thoughts and philosophy, the more interesting the conversation will become; the more open the team is, the more likely information will be comfortably shared. You do not want anyone afraid to bring forth ideas or problems to the group or at least to other team members individually.

You may consider writing a vision statement for the program. If so, it would be best to find those interested in "volunteering" to do so. As a leader-manager type, you may want to facilitate this activity yourself, but it could be the perfect time to help budding leaders on the program facilitate this session. As a good leader, you should always look for ways to step back and help others lead. The results may not be just what you wanted or expected—but, you know what—what you wanted or expected may not be as good as what the team wants! Always fight Arrogance—the disguised Impersonator of Leadership Confidence.

Some companies use "team-building exercises" or trips or activities to build the team culture. My experience is that these are fun and filled with learning, but really do little to build teamwork. What builds a team is a recognized common goal and the desire to help everyone on the team to achieve that goal—with a true interest in the personal wellness and happiness of each teammate. When someone is sick, the team should actually care about the person, as well as the person's tasks. You set the example for the team. I do not believe you can be an effective PM unless you can be an effective leader, and I do not think you can be an effective leader without caring for the people on the team.

When my children were little, my wife was injured in a car accident. I had limited vacation time, and no one to help my wife heal and take care of the children while she did. My boss said simply, "Take care of your family—do not worry about work or vacation time, we will work that out." In addition to giving me relief from my immediate concerns, it made me really appreciate the company (who my boss represented). For years, and, in fact, to this day, this example of putting me ahead of his work concerns has paid dividends 100-fold. Caring leaders are the most effective in getting the work done.

END OF CHAPTER QUESTIONS

For Discussion

1. Discuss the importance of allocations and some of the issues around making "contracts" with the functional departments.
2. What are some of the considerations in selecting/appointing the sub-team leaders?
3. Sometimes it may be positive to outsource part of your product. What are some of the considerations if you would like to do so? Are other stakeholders affected? How can you make outsourcing a product more likely to succeed?
4. Do team-building exercises work? Are they always worth the expense? How can you design an exercise that is effective?
5. Discuss what "trust" means as a leadership value. What are some strategies to building a trusting culture, and what are some behaviors that can ruin that culture?
6. We are running a business here!
    a) We do not have time to care about people!
    b) The health and happiness of our people is the basis of our success! Do you have any stories to illustrate theory a or theory b?

Written Assignments

1. PMs are both managers and leaders. Describe how both roles come into play at the start of a program.
2. Why is the selection of the team so important?
3. What are some ways to establish a shared vision for the program?
4. What are some of the advantages and disadvantages of utilizing contract engineers? Tell us a story about some experiences you may have had with one or more of them. Or, if you do not have that experience, tell us how you would select a good contract engineer.

7

Running the Program

At this point, let us assume that you have completed the basic planning for the program—there is a baselined schedule; a spend plan for each budget center (engineering, contracts, quality, etc.); you have organized your program documents in a way that they are accessible to everyone on the program; you have created, either formally or informally, a vision for the program; you have established regular team meetings; you have your core staff in place; you have met with the customer; you have established your risk register and put aside reserve funds to manage those risks; you have invited customers at every reasonable level of management to visit your plant for a kick-off meeting; you have established a schedule for customer phone calls—ideally, weekly.

With all this preparation, it seems that the program will virtually run itself!

But, let me assure you, that even if you do all those things perfectly (and you will not), things will not go according to plan.

Knowing this, and anticipating this, is a very important part of the program manager's (PM's) job. There is a common belief that one must "stick to the plan." While following the plan is valid to some extent, the effective PM is always alert to emerging issues that require adjustments to that basic plan. You could lose your key systems engineer; the customer could ask for a proposal for additional scope; or your design team discovers that the 15 board styles they conceived are not enough—that 17 boards will have to be designed (where will you get this funding?). Recognizing when to "stick to the plan" and when to change the plan is what you get paid for and is your responsibility.

This is one of the main areas where that Venn diagram overlap of management and leadership occurs. You must not allow problems that should be solved as part of the job become cause for a relaxation in schedule or the allocation of reserve funds. Problems are a natural occurrence on the program, and most of them should not cause a re-plan. If you do not require progress on the plan but allow unsolved problems to change your schedule demands, you will run out of time and money way before the job is done. Enforcing schedule performance is the management part of your role. But *how* you "demand" that performance is the leadership part of your job.

LEADERSHIP STYLES

Leadership is vital to the success of your program. Perhaps because it is an esoteric concept, or perhaps because there is such a wide range of behaviors that are taken to be "leadership," the concept is generally poorly understood and often poorly implemented. It is so important in program management, that it warrants a separate discussion in Chapter 9. For now, we will see how two different styles affect and effect the performance of your team.

The absolute best way to provide the proper impetus and challenge to your team is to emphasize the "we are all in this together" view of the program. This is a relationship-based leadership style that makes your colleagues into your collaborators rather than your adversaries. Old-school thinking might approach this issue as "You made a commitment to get that board designed by March 15, and you darn well (or stronger adverb) need to meet it!" Notice the implied threat. Sometimes it's not even implied: "… or I'm going to your boss," "or you are off the program," "or I will humiliate you in the community."

If you make your lieutenants fear you, you can be sure that they will not tell you about their concerns. If you make them defensive, you can be sure that they will spend their time protecting themselves rather than doing the work. To make "holding someone accountable" your basic style of management is a dangerous as well as simplistic model. Sure, you want people to do their best to control costs and schedule. But it is so rarely the case that they are not trying hard. Therefore this tool—"hold-them-accountable" style—that is intended to make them try harder is just the wrong tool. People naturally want to succeed and want the program to succeed. If you want people to work together to solve problems, you must make it in their personal best interest to work together.

Here are two scenarios that illustrate this point:

Scenario A—Unenlightened (Old-School) Program Manager

Lead Electrical Engineer (LEE): Bill, it looks like the Framazoid Board
        design is going to be two weeks late.
Unenlightened Program Manager (UPM): WHAT!
LEE: Yep.
UPM: Just wait a minute, Mary. You committed to having that board
        designed by June 1. You just better find a way to get it done.
LEE: But don't you remember, the customer pointed out the obscure
        requirement that each part had to be de-rated by an additional
        10%, on that board, and you acquiesced.
UPM: Don't give me excuses. Give me results.
LEE: But…
UPM: No "buts"! I want a recovery plan on my desk tomorrow.
LEE: But…
UPM: I told you no "buts"!
LEE: Ok, I'll try to come up with something.
UPM: You better do more than just try. And, I want to be briefed daily
        about your progress against your recovery plan. And I want that
        report written so I can keep an eye on you.
LEE: Yes, sir.
(LEE leaves.)

Now your lead electrical engineer (EE) has two problems—getting the board designed and managing your anger. She has to come up with a new plan (which is probably just lip service) and report on it. And now her concentration is focused on building a case as to why she is not incompetent, but why *you* are incompetent. She may spend some time documenting the fact that you gave in to the customer and caused an additional scope change to slip into the contract, which even if it were not true, will take time to refute. She is likely to go to her functional manager (FM) for protection of her reputation, and he or she will have to be briefed, and will probably align against you. She will have to tell her team that a recovery plan has been mandated—and she may resort to some reactionary behavior, such as the dreaded "mandatory overtime." What a mess!

But the UPM may be thinking: "By being forceful and demanding, I solved the problem! Mary is making a recovery plan and the problem is solved." This is a view of reality obscured by the PM's arrogance. You can see the problem just got worse. Mary is not even working on solving the problem anymore—she is working on her image and plotting against you!

And, even worse from a strategic point of view, do you think Mary is going to come to you with problems ever again? No. She is not stupid—the last time she brought a problem to you, not only did you not help, but you also gave her other work (a formalized recovery plan and reports) and caused her to perform distracting "work"—documenting that the problem was not hers but yours.

Here is how a more Enlightened PM might handle the same situation.

Scenario B—Enlightened Program Manager

> *LEE:* Bill, it looks like the Framazoid Board design is going to be two weeks late.
> *Enlightened Program Manager (EPM):* Oh, no. What are we going to do?
> *LEE:* Well, maybe we could talk to Fred who's designing the board and see what the problem is.
> *EPM:* That sounds good. Let's call him.
> *Fred:* I'm having problems with the additional de-ratings we have to do, and it's just taking more time than I expected.
> *EPM:* Can we get someone to help?
> *Fred:* I think our engineering aide, Ann, could do the calculations for me. It might be a stretch, but I think it would be good for her to learn anyway.
> *EPM:* What about the extra hours? That will add cost to the board design.
> *LEE:* Good news, there, Bill. Ann is indirect—she charges budget because she helps out on many programs.
> *EPM:* Sounds good, but can we get her?
> *LEE:* I think so—I have good rapport with her boss and he's a good guy. I think he can make it happen—but I will check.
> *EPM:* Sounds like a plan. Let's do it!

Well, that is quite a different outcome. Sure, it was somewhat fictionalized and simplified, but it illustrates the principle—if you help your team with their problems rather than just applying pressure, your team will work with you to minimize problems as soon as they appear. You will aim their creativity and initiative toward solving the problem. There is no reason to conceal problems from you, because they see you as helpful rather than threatening. The EPM reflects on that encounter and does not think "By being forceful and demanding I have solved the problem." He thinks, "Whew, that scared me for a minute—it's a good thing we have smart people on this job!" It is not all about him; it is all about "us."

Not all scenarios are this straightforward. Oftentimes, problems are not resolved in one let's-work-together meeting. But, the more complex the problem, the more you need the team working together, <u>with</u> you (not <u>for</u> you) to solve the problem. As the PM, you, more than anyone else, have the opportunity to establish the culture of the program. By setting the expectation that smart, hard-working people, by working together, will succeed, you have greatly increased your chance for success. Building teamwork—by leveraging the instinctive behavior that has moved humans to the top of the food chain—is easier than one might first think. The hunting party that communicated among itself and saw the goal of killing the dangerous beast for food for the tribe, is the very model on which you want to build your team. The more you hear "we" when your team gets together (and the less you hear "I" or "he" or "she"), the closer you are to using that millennia-old conditioning. The hunting party that worked together succeeded and went on to breed new cooperative, communicating hunters. The hunting parties that failed to work together naturally selected themselves out of the gene pool. Use that inherent team-bonding behavior to your program's advantage.

Thus the most effective "management style" is the zealous promotion of "we are all in this together—we succeed or fail as a team."

One enlightened senior manager once told me that she was tolerant of a range of "management styles"—as long as they are successful. Perhaps we should consider that as semi-enlightened thought. Actually, this comment was intended to allow more people-centric styles to be tolerated—at the time the local culture was very much an intimidation culture. Over the years, there has been a true movement toward "softer" styles, and this senior manager's comment was an indication of tolerance of those styles, which until then may have been considered ineffective.

Is there a reason that the non-intimidation style of leadership has become more popular over the last 20 years? One theory is that our relationships to our managers are based on relationship models learned in our childhood with our parents and teachers. A generation ago, it was usual for "discipline" to be corporal (it was OK to spank your child, or your student, even in public—can you believe that?). Or hear "I'll give you something to cry about!" But in these last one or two generations, corporal punishment of children is illegal in schools and much less common at home. If you are raised in an intimidating environment, well, intimidation in the workplace is tolerated, if not even expected. However, if you are raised in an environment that promotes personal responsibility, then cultures that expect personal responsibility will be effective.

But the senior manager has the responsibility to go beyond "tolerating" more effective leadership styles—for example, she has the responsibility to coach her under-managers to use those more effective styles. That's why I consider this comment only semi-enlightened. As leaders we have the responsibility to help others do their jobs more effectively, and this expressly includes other leaders who report to us or work with us. By helping them be more effective, you make the organization more effective.

The development of team concept and team identity is vital to this style of leadership. That is why we have said quite a bit in earlier chapters on selection of the team members. Having a "good team" is the very most important way to ensure your program's success. Previously, we have discussed the very important quality of selecting people who have mutual respect, or at the very least, do not have animosity toward one another. And, we have discussed the importance of selecting team members who are technically competent for the tasks that they will be assigned. Our emphasis has been the importance of selecting members that are hardworking and of high integrity. But here is one other important trait that you should look for: self-confidence, which also promotes nicely into team-confidence. This "we will succeed" expectation is the self-fulfilling prophecy you want in your team. For extra credit, if you and they can generate an "enthusiasm for success" on the program, your leadership job is half done. One of the worst things you can hear from your team is "well, we will do the best we can." This indicates an expectation of failure. If you hear this, you should ask why they lack the confidence to succeed—and then solve that fear. You must change the "we will do the best we can" to "it is going to be tough, but we can do it!" And let us not forget that you as leader of the team are also a team member. You need people who will build *your* confidence when things get tough.

MAKING PROGRESS AND MONITORING PROGRESS

Progress and its measurement are the key responsibilities of a PM. There is no question as to the centrality of this function. But there is a wide range of how it can be, or should be, achieved.

The first step is having a good schedule for your program. It is your roadmap of how to get to the end goal. A detailed schedule for a 40-person, 4-year program will run several pages wide and several pages deep. So much information will be there that you will have a hard time seeing the forest for the trees. And creating this detailed schedule is a monumental and time-consuming effort. And suppose you create this very detailed schedule—who will ever spend the hours necessary to study it? And 3 months into the 4-year program, something will change, and the schedule will become obsolete.

So, should you not make (or have made) a schedule? Of course not. The best approach to this issue is to create a top-level schedule for the whole program—one that will capture the vision of the program and will serve as general direction. Detailed day-to-day schedules are also needed, but need only span a period of weeks or a couple of months. This is "rolling wave" scheduling, wherein only the next few weeks or months are scheduled in detail, and of course must fit into the overall schedule. These detailed schedules should involve the people doing the work as much as possible—but remember, not everyone is good at this type of planning. What you want is worker-level buy-in to the schedule. You may be working through lead engineers or FMs, and if you have the right lieutenants, they will ensure the workers' buy-in. But do not assume they have done their sub-schedules collaboratively—because it is easier not to do so. Of all the skills on your team, leadership is the one that you are best equipped to coach. So ask your lieutenants about schedule buy-in of the workers and how they achieved it. If they look at you funny, you should worry. If they say "I told them what they have to do, or else," you may not be building the culture you need.

And when your sub-schedules are created, one thing is very important: they should have your people's *names* on the tasks you need them to do. I have seen detailed schedules ignored by team-members, until you put a name on the task. People will actually scan for their name when handed a schedule. Making clear what has to be done and by whom is job one for you—and making it clear to the "whom" that has to do it is key.

So now, let us assume you have these detailed schedules. Now you need to be vigilant as to the progress to those schedules. Almost always, things tend to take a little longer than planned. And for engineers who are driven to do excellent work, there is a tendency to polish the work to the extreme. This "polishing" is one of your biggest schedule risks. Here is an analogy that might help you help your folks understand the scheduling imperative.

Suppose you are painting your bedroom. There is no other place for you and your spouse to sleep at the end of the day; you start early, so you will have plenty of time to do the job right. You start with the windows—there are three. The first one takes one and a half hours, and you realize that you better pick up the pace—ok, let's not mask the wall, let's paint a straight line. Now it's time for lunch—hmm, better not go out; better make a sandwich and save time. Hey, the ceiling needs some patching, and masking around the border. Now it is 2:30 p.m., and you have not painted the trim. Better freehand that too. It will not be perfect, but you have to sleep there tonight. OK, now it's time to get to the walls, but, gasp, it is almost dark—and you will need to bring in more lighting. Going out for dinner is out of the question—maybe just an apple or two. And you work to get the room finished. Figure 7.1 shows the work about to begin, and Figure 7.2 shows how you would self-manage your work to get it done on time. The transference of the schedule concern from you, the PM, to the engineers performing the work is a key accomplishment for you to achieve. It has to be their commitment to their own schedule.

Your engineers might feel like they need to mask the molding—but really they do not. They have to feel the schedule pressure in the same way as they would about finishing the painting. If you look closely, the freehanded

FIGURE 7.1
Time-sensitive task.

(From Dollar Photo Club, File #42649342 Dayanadesign.)

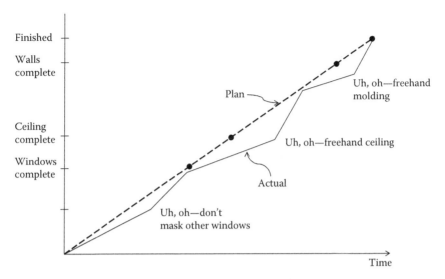

FIGURE 7.2
Schedule self-management.

painting might look just a little less perfect—but in reality it is very much acceptable. The key is to coach your team to think of the schedule in this way. They must themselves figure how to expedite the work to fit the schedule while preserving the adequate (not perfect) effort. Remember, "perfect" is the enemy of "excellent"—and "perfect" is the *mortal* enemy of your schedule.

MONITORING PROGRESS—METRICS

As you created your detailed schedule (top level for the whole program, detailed level for near term work), you identified tasks and how much cost (or, for design programs, engineering hours) is allotted to each task. For large programs, often you will have the help of a Scheduler, and perhaps a financial analyst (FA) to help you manage the program. (I say "manage" because they will help you *manage* the program, but will be little or no help in *leading* the program.) The tasks and hours represent budgeted work (BW). As you complete tasks, you create actual work (AW). By quantitatively calculating the actual completion of tasks and comparing it against the plan, and by comparing your scheduled work dates against your actual completion dates, you can do several very significant comparisons and calculations. Following here is a brief overview of earned value (EV) analysis. A more detailed discussion, showing examples, is provided in Chapter 11.

EV—This is the dollar value of the tasks that have been completed. You would want to have an EV that is greater than your actual costs—indicating that you have accomplished more than you planned for the money you spent. The EV concept is used in creating the cost performance index (CPI), which is calculated by the formula:

$$CPI = BCWP/ACWP$$

where BCWP = budgeted cost of work performed and ACWP = actual cost of work performed.

Clearly, you would want to have a CPI that is greater than unity—more work accomplished for less money than planned.

A similar calculation can be performed with respect to scheduled tasks, the schedule performance index (SPI):

$$SPI = BCWS/BCWP$$

where BCWS = budgeted cost of work scheduled and BCWP = budgeted cost of work performed.

The SPI allows you to see quantitatively if your work is being done ahead of or behind your planned schedule. Notice that it does not use the cost of that performance, it is just the comparison of how much work you have scheduled to have completed against the actual work that has been completed. As with the CPI, you would want your SPI to be greater than unity.

So by monitoring your CPI and SPI, it would seem that you could be sure your program was on cost and on schedule. Yes, it would seem that way. The mathematics of this comparison are straightforward and perhaps comforting to the analytical mind. But there is one important flaw in using these indices blindly: they rely on a very accurate appraisal of the work being captured in the schedule. For example, if you inadvertently overvalue (in dollars and schedule time) some of the early tasks on a program, you can seemingly be doing a great job against your CPI and SPI but still be heading for trouble.

The EV concept is thus best applied for production programs, where you may have actual history to determine how long and how much it takes to build, for example, certain assemblies. This history makes for an excellent basis for your schedule. However, where design work is concerned, it is much more difficult to estimate how long it will take to create a design. There are many unknowns in creative work, and your ability to estimate them is probably less inherently "correct." In fact, if you think about it, the actual time that it takes to do the work is really much more "correct" than the estimates you made of the work when you created the schedule, or quoted the program. That being the case, the EV approach might better be seen as a measure of your ability to estimate the work, rather than the efficiency of how the work is actually being performed. But philosophically interesting though that may be, that is not the generally accepted belief. Blind use of EV analysis can lead to a false sense of security early in the program, especially as the early tasks are generally lower risk. For example, creating a requirements matrix can be considered "done" when it needs to be done—the variable is how complete it is. But a circuit board, designed to those requirements, actually has to work—and that is plagued with schedule risks. So your CPI and SPI are only as good as the crystal ball that you used when you created your schedule, and calibration of a crystal ball is not very easy! See Figure 7.3. And a favorable SPI may be comforting, but unless you are keeping an eye on the critical path, that favorable SPI can lead to an unfortunate surprise—most of the program could be ahead of schedule, but the pacing activities could be in trouble.

FIGURE 7.3
Not the best measurement tool.

(From Dollar Photo Club, File #28487986 koya979.)

Therefore, use the PM quantitative tools wisely, putting the appropriate reliance on what they are telling you based on the quality of the schedule you are using. Appropriately used, especially for very knowable applications (such as production runs with good historical basis), these tools can be helpful.

FOCUSING ON QUALITY

While you are monitoring your team's progress, both practically and numerically, you must continually monitor the quality of the work, and, possibly more importantly, the customer's perception of the quality of the work. Think of the times that you have bought an expensive product—a car or a TV, for example—and suppose it failed to perform as you expected. Your loyalty to that brand evaporated—and perhaps never recovered.

So it is for the future of your business with your important customers that they must be satisfied with your work. Traditional customers are probably the largest source of future business, and you and your company cannot afford to lose their loyalty to your company. Your team, therefore, has a larger-than-program responsibility to satisfy the customer. High-quality products—reports, drawings, models, hardware, and software—are the key to building not only "brand loyalty" but also customer rapport.

That rapport, built by respect for the work that you and your team do, is key to success on the program. There will be times when things do not go as planned—your customer's willingness to tolerate setbacks and work with you to minimize consequences will be important. You will not get concessions over gray requirements or reach agreement on scope issues if your customer is displeased with your work.

As the leader of the program and as the chief interface to the customer, you must see the quality and perceived quality of your work as your personal responsibility. While customers may vary greatly in their expectations of what constitutes high-quality work, you should strive to please the most demanding of customers. This means significant items such as making sure the specification requirements are met—and smaller ones such providing a final proofread on reports for grammar and spelling. A technically excellent report can be "wasted" if it contains even a few grammatical errors or editing mistakes. Take the time (or rather plan for it) so that a final review (and corrections) can be made of submittal documents.

And of course, delivered hardware and software has to be flawless. Many of the systems that you may work on are depended on for the safety of the user. Thorough testing is required to ensure that the equipment conforms to the specification. Furthermore, you should seek out engineers, operators, and technicians who have a true affinity for quality. No test specification can be written flawlessly to identify all possible problems. You want a test technician who has a questioning and maybe even a skeptical attitude. If you specify that the equipment should present a certain output at a certain input, you certainly want the technician to verify this is true. But if a circuit card gets noticeably hot, even while working, and even though temperature of the board is not in the test specification (T-spec) as a requirement, you want someone who will notice it and bring it to the attention of the designers. This behavior represents a quality culture that, hopefully, your company has. If it is not strong, it is incumbent upon you to create that culture in your program. You can do this in several ways, but one of the most effective is through communication. You should get the team together and thank them for their good work and brief them on the importance of the mission of the product and the importance of high quality to the user and to the customer. They must know that when something does not look right, you, as the PM, want to know and investigate it. If you do not explicitly say so, there may be a belief that your schedule affinity may be greater than your quality affinity, and testing "anomalies" can be ignored if not specifically called out by the T-spec.

For the design aspects of the program, you should insist on meaningful design reviews, attended by critical, demanding, senior engineers. And you should encourage informal peer consultation and peer review as the design work is taking shape. While design reviews are important, and probably mandated by your company's procedures, they will not catch all errors. That is why the establishment of a "we are all in this together" atmosphere is so important. If your engineers review their work with their peers, you will build a learning culture as well as a quality culture. Resist the urge to worry about the costs of reviews and collaboration—these costs are small compared to the costs of a dissatisfied customer. A zealous drive to high quality is your best trait that leads to a happy customer.

With any customer, but especially with an on-going customer, your immediate attention to problems with fielded equipment is vital. When your customer deploys equipment to the field, his or her reputation is on the line as well as yours. Problem reports need prompt attention. Solving these field problems can also be important learning experiences and quality-culture drivers for your engineers. And if a problem is not solved immediately, your constancy in resolving the problem is vitally important. Send in an additional team if necessary to help one that is "stuck" on a problem. It is hoped you have money in your program for this support, but it is vitally important that you act as soon as possible. If there is some question whether it is your equipment at fault or a piece of interfacing equipment, respond quickly nonetheless, to either find that your equipment is satisfactory or that it has a previously unrecognized flaw. Many customers are willing to have you go to the field on troubleshooting missions with a "loser pays" agreement—if the error is in your equipment, it is a warranty charge—if it is an installation error or malfunctioning interface equipment, then you will be reimbursed for your efforts. Follow up on even successful missions to enhance your learning and to show that you care about the equipment's performance.

Of comparable, if not even greater importance than the quality of your work and equipment, is the customer's perception of it. Of course, the best way to have the customer think that your work is good is to produce good work! But that may not be quite enough. Misconceptions can arise. One of the best ways to prevent this problem is to have periodic (for example, weekly) calls with the customer. It is best to schedule these for a particular day and time each week. Actions for both you and the customer can be identified—the faster you close out your actions, the better you will be perceived. These action items agreed to can be tracked with formal of semi-formal minutes. This will prevent minor misunderstandings from going undetected until they become major problems.

Some companies require, or essentially require, customers to provide formal ratings on the program's progress from their viewpoint. Many customers see this as an opportunity, but some see it as providing them additional personal risk—what if they give you good ratings one month, but the next month a major problem is unearthed? This could be very embarrassing to the customer and make the customer look uninformed to her or his management. The formality of the report therefore could be a reason for him or her not to want to provide that feedback—so work with your customer to a solution that is comfortable and meaningful for both of you. Remember, you are seeking to understand what your customer thinks of your work—that can be very informal and just between the customer's lead and you. Your management may want to know what the customer thinks, but should rely on you to provide unvarnished feedback on what your customer tells you.

MANAGING THE CUSTOMER

One of the best ways to ensure that your work is being well received is by providing a written progress report on a monthly basis and asking for feedback on the report. It provides a vehicle for communication and allows for any potential disagreement on scope, quality, or schedule to be identified as early as possible. Furthermore, it allows you to put "on the record" information that could be very helpful if things happen that are not expected on the program. Complex programs can certainly run into problems, and the fact that you have identified the concern early and are taking steps to resolve the concern can be very helpful. And if your amiable customer gets reassigned and a more curmudgeonly one arrives to take her or his place, you have a documented history of good performance to set positive expectations. If you, or a member of your team, write the monthly report, it also gives you control over the "record." Missing or delayed information can be identified as soon as it is discovered to be missing. You and the customer will have a record of when the information was requested, in the case that it takes an undesirable amount of time to get your answers. And it can be repeated each month. Never underestimate the power of persistence, especially if it is documented.

The monthly progress report also allows a semi-formal notification to the customer of issues that need to be resolved, especially ones that requires the customer to take action. The semi-formal report, which will be read by your direct customer and probably a few others at either your customer's office or at yours, is a great way to ask for information that you need. This vehicle is one step above a phone call or e-mail, but is one large step below a contractual letter. The contract letter will get plenty of attention, being read by administrators at your organization and your customer. If your customer is late in, for example, taking approval action on one of your submittals, certainly you have a contractual right to "blow the whistle" on him or her and "make him or her perform." But, think about it; if you do that you will be impugning the performance of your customer and, being human, he or she will resent it—even if, or maybe especially if, they are negligent on this item. If you formally document, through a contract letter, a deficiency on the customer's part, you better be ready for greater scrutiny in the future. Thus, you are playing with fire.

Of course, sometimes it is necessary to formalize your complaint about missing information or delayed customer action, but the process that has worked for me is a gradual escalation of the issue:

1. Verbal request—maybe a couple or a few times (friendly tone)
2. E-mail request—again, more than once (more friendly tone)
3. Entry of the need into the progress report (semi-innocuous)
4. Entry of the need into the progress report in a "Critical Items" section
5. A formal contract letter stating the need
   a. Note that this may provoke a need to identify the impact on the program of the delayed information. You may need to provide an estimate of the cost and/or schedule impact. This will go rapidly up the customer's chain, as it is approaching a "claim."
   b. If it gets this far, there is a great chance that all the efforts you have spent to make a good working relationship with the customer will be lost, and you will enter into an acrimonious phase of customer relations. If this happens, and it is severe enough, it is probably best for the program to change either the customer representative, or you, the PM. (The program will not be successful if the customer is angry with the PM!)

As you can see, the best way to handle problems is at the relationship level. In fact, if you and your customer can form an alliance to make progress by helping each other, you are doing a great job in customer management. In the example above, while you are trying to get the information or the approval you need, let your customer know that you are doing all you can to avoid escalating the concern to a higher state. If you do this sincerely, not only will the customer focus on your need, but by refraining from "blowing the whistle," you will be deepening your collaborative relationship, and when you have a problem, he or she will treat you with the same consideration. Remember, in the power balance between you and the customer, you have 10% and the customer has 90%. No one needs a powerful enemy, but everyone needs a powerful friend!

As a warning, however, do not let this relationship, although very important, cause you to adversely affect your program's progress or put it at an unreasonable risk. You are playing with fire here too, if you "conceal" customer deficiencies that can cause you to spend time or money to work around delays. Remember that your primary mission is to get the job done, with high quality, on schedule and within budget. Good customer relations are a tool toward this end. But being overly tolerant of delayed customer performance items is one path to failure that you must avoid. You do not want to have to explain that you missed a ship date because the customer failed to approve a document within the allotted span and you took no (formal) action to resolve the problem. If that happens, neither you nor the customer you hoped to protect will benefit.

IDENTIFYING AND AVOIDING PERFORMANCE TRAPS

One of the most important ways that you can fulfill your obligation to get the work done on time and within budget is to be vigilant and quickly identify when things are going off course. On design programs, this is particularly important—and particularly challenging. In general, it is great to work with engineers—through a combination of personality, schooling, culture, and intelligence, they, as a group, have a very high work ethic. But while you may not find your engineers appearing to be relaxing, and would rather find them working, it is very possible that you will find them working on the wrong thing, or stuck in some unimportant issue that does not affect the program.

This is another area where your management ("Get it done!") and your leadership ("What can I do to help?") need to work together, with your perception, to make progress and keep your team engaged. One way to do this is to ask plenty of questions—"Why is it necessary to get that Spice (a modeling program) to correctly predict the circuit's performance?" You may find that, in fact, it is not necessary—that the circuit is built up and working just fine, it is just the engineer's sense of closure that he or she wants to spend time refining/correcting the model. If the model was just to be a design aid, then, for your program's interest, it is only a moot point that the model has a flaw.

But here is the challenge—once you have identified that the engineer is working on something that he or she thinks is important, and you, in your wisdom, do not think is important, action is needed. And it has to be an action that does not disengage your engineer from the project. If you say, "Mary, the Spice model is not important. Stop working on it," you probably will get her to stop, but you probably will not get her to agree. The more you manage by dictum, the more you own the work and the less the engineer owns the work. A better approach is to discuss with Mary why she thinks it is important to refine the model. In fact, you may be wrong, and there is a need. But as you talk through it with her, and as you explain the schedule and cost pressures that you (and she!) are under, it is likely that she will agree to put the modeling work aside and move forward with the next design task. This conversation gets the work on track and keeps Mary working on what both you and she agree is the most important thing to do next.

In some organizations and on some programs and projects, you may be buffered from this level of detail by your lieutenants—either FMs or lead technical people. But this is not a task that you can delegate blindly. FMs are interested in their engineers learning new skills and growing technically. Lead technical folks on your team may also lean toward technical purity over getting things done. And if you have dictatorial lieutenants, you may find your engineers "just doing what they were told" and leaving at 5:00 p.m. no matter what the problem is. While undue invasiveness of the PM must be avoided, this area is so important that your monitoring of it is absolutely not invasive!

Of course, if one of the engineers wants to pursue details that you both agree are not necessary for the program's forward motion, he or she can certainly pursue that learning experience on his or her own time. In fact, this is admirable from a professional development standpoint. You should commend them for doing so and let their FM know that they are doing it on their own. It is very reasonable to expect that high-performing professionals will devote significant time to their own professional growth, and if the area of their interest is at least close to what your program needs in terms of skill base (in our example above, Spice modeling) that is a triple benefit: to the engineer, to the company and to the program. But of course, it is not proper to allow these nonscope, but adjacent tasks, to be charged to your program.

Another solution to this age-old dilemma is to assign work that aligns with the engineer's particular strengths.

This is an example of a lesson in leadership that I learned from one of my early managers. I was an inexperienced EE working on a program and was assigned to bring a design to completion. Also working on this project was a much more senior (and much more gifted) engineer. Let's call him John. But my boss assigned me to complete the design by a certain date. John completely understood the technical issues and completely understood the solution. In fact, he could envision multiple solutions and was deeply interested in investigating all of them. He was so interested in learning more, that he had no interest in getting the drawings completed. (Of course, this continual quest for learning is what made him so talented.) However, John lacked the "completion gene." My boss was smart enough to assign me to complete the drawing (he knew that I had the completion gene), and he knew that I would be able to focus John, if only out of pity for my dilemma of having to complete work I did not understand, on getting the design done. Now at the time, it was not easy for me—John wanted to teach me in detail all of the possible solutions. I was interested of course in learning, but possessing the completion gene, I wanted to pick one and finish the design. Learning could wait, in my opinion, for after we had the drawing into drafting!

So it was a little backward, having the junior person responsible for the work, while the senior person knew the design. But it worked! By properly identifying the talents and nature of the engineers, my boss was able to get the work done on time, while John could continue learning, and I, as it turned out, got some very valuable experience in getting things to closure, without having to resort to using power—which, of course, I lacked! Now that's just smart delegation!

This example and the discussion above should make it clear that as the PM you must always be alert to problems and distractions that are compromising progress. Although every situation is different, the solution is always basically the same. Work to understand what the people on the project are working toward, attempt to determine if it is on course and necessary, and discuss with them the value of the task they are working on and how it fits into the overall schedule and goal. Always, always, expect them to know more than you about the work they are doing—because they obviously will. And always be open to understanding what they think is important. You will need to trust them (and tell them that you do), once they understand the cost and schedule constraints, to only concentrate on things that are necessary for the program.

GETTING "STUCK" AND GETTING "UNSTUCK"

At times during the course of a design project, your engineers may become stalled on a technical issue. It is your job (although you might be assisted by technical leads or functional management) to identify that someone is in fact stuck and to help them get unstuck. You should be vigilant enough to notice if something on your program is taking longer than it should. Even if you have a task assigned, assume for example that it is 2-month long, you would not want to wait until the end of that 2-month period to realize that only one day had been earned! (This is yet another example of how you cannot trust your metrics to adequately gauge your progress.) This is a particularly sensitive but very important task for you—the unsticking of stuck engineers. It is challenging because very often the engineer truly believes that he or she is on the verge of a breakthrough. You do not want to improperly "distrust" them, but you cannot wait for a breakthrough that does not happen. Knowing when to intervene is thus the challenge, and knowing how to intervene is the next challenge. Similar to many of the problems we discuss here, the solution is rooted in much the same philosophy—treat people with respect, being humble enough to realize that they know more than you do about the technical problem, make it clear to them the schedule and cost pressures you are facing together, and work with them toward a solution. The classic program management question— "Is it done yet?"—is helpful in this situation, but the effective PM goes one step further. That one step further could be increasing the frequency of the "Is it done yet?" question, asking it with an edge in your voice, or asking the FM—but hopefully you can do something that is more effective. By discussing the problem with the engineer, you and he or she might agree on a course of action, by which you actually may help the engineer! This transaction, where you decide together to, for example, bring in an expert to consult, is founded on your approachability and problem-solving-constructive demeanor. No one will discuss problems with you if your only response is to get angry and apply more pressure. Your team must continue to view you as someone who is approachable—and consistently so.

If Engineer A sees you yelling at Engineer B, Engineer A and the other team members that see or hear of the incident will avoid telling you things that might make you holler! Cut off from information on problems, you must use guile or intimidation to get the information you seek. And remember, engineers are smart—they will easily be able to manage the news. But you do not want them managing the news; you want them working on the problem.

So the challenge is to welcome good and bad news and work as a trusted colleague to help solve the problem. Start off by assuming that an engineer's ego is at stake if you suggest bringing in an expert to help. (Think of yourself in that position and it is easy to see the risk.) But if you and the engineer talk about what the challenge is, it could easily be his or her idea to find help. This way, the solution is in the ownership sphere of the engineer with the problem. This remains the challenge: keeping the engineer owning the problem and not being afraid to seek outside help. Always use positive reinforcement—"That was very smart of you to get Mary to help with that problem—some engineers would have beaten their heads against the wall for days before asking for help." Contrast that with: "What's the matter with you? Don't you know anything about Op Amps? And you call yourself an engineer?" Trust me—I have seen both approaches, and there is no comparison as to their effectiveness.

CUSTOMERS AS MOTIVATORS

Most of us have an innate desire to please our customers—it is some combination of recognition that customer satisfaction is the prerequisite to continued employment and the desire to please "parent figures." Whatever the cause, you, as the PM, can use the customer to help motivate your team. The more you can make the customer's appreciation of their work known to the engineers, the more effort and "heart" they will put into their work. In fact, once they have established a reputation with the customer as being talented and high performing, they will go to great lengths to ensure that their reputation is not diminished.

One of the best drivers of progress is the customer demonstration. Dreaded as well as loved, demonstrations and meetings are included by smart customers in procurements to not only measure progress but to encourage it as well. The establishment of clear goals is extremely important to focus your team on progress. And while you as the PM may be very capable of establishing these goals, they will be recognized as internal goals. Customer demonstrations are thus inherently different. Whereas someone on the team might decide your self-established goals are not relevant, important or necessary (of course you will remind them that they are), no one on the team would even bother to question the necessity of demonstrations that are in the contract. A customer visit is important and will also likely be on the radar of senior management. So you will have to have all the motivational forces acting in unison:

- Dedication to the team's success
- Desire to make the program successful

- Desire to please the customer
- Desire to enhance one's reputation with the customer
- Desire to enhance one's reputation with colleagues
- Desire to enhance one's reputation with FMs (raises)
- And desire to please you (the PM) and keep you out of difficulty with your management and the customer

Coupled with a clear goal—"We need to get this subsystem working by April 14!"—great progress will be made. Unnecessary work (the refinement of the Spice model in the example above) will automatically be put aside. The team will come together as those not directly involved will put their tasks aside temporarily to make the demonstration a success. People will work late (and you will bring them pizza) so that the goal can be met. Driving successfully to excel during this demonstration will, all by itself add to the motivation of the team that will last well beyond the demonstration itself.

Keep the power of the customer demonstration in mind as you work with the customer in formulating the details of the contract. While inclusion of a demonstration seems like a risk-increaser (because it is), you may very well want to agree to it or even suggest it for the motivational value it provides. And although, as the demonstration approaches, you may wonder what you were thinking, the benefits will become obvious in the work tempo and demeanor of your team. It is just plain exciting to be able to (and even to be forced to) show off your work to an audience.

To a lesser extent, the same is true for deliverables on the design phase of the contract. These tend to be less visible and less team-wide than a demonstration, but they also have motivational and focusing capital for you. Customer dates are known and tracked by folks outside your program and folks that are higher up in the management hierarchy. A program rich in customer deliverables is therefore positive, although again it provides an increased level of stress for you and the team. But that stress (or rather the relief of stress) is an important motivator.

One thing to be careful of, however, is to ensure that the deliverables are appropriate to drive and or monitor the program's progress. Unimportant reports or analyses can be a huge distraction from the actual work of creating a design. And again, your ability to work with your customer as the deliverables list is being formulated is very important. In some cases, you may not have access to the customer at the time the list is formulated—it may be, for example, part of the bid package. But nonetheless, even if it is in the contract, you and your customer will likely be able to revise the list to something more meaningful—once you have established the customer's confidence that you are trying to help the program move forward, not trying to get out of work.

KEEPING SENIOR MANAGEMENT ENGAGED

As the PM you will be very busy—keeping the program moving, monitoring progress, monitoring costs, motivating the team, satisfying and building rapport with the customer, achieving deliverables, AND making sure your senior management is confident that the program is on track (and that you are doing a good job).

How your senior managers measure your progress varies from company to company and from manager to manager. It varies over time, also, as the company's culture evolves and as lessons are learned from successful and unsuccessful programs. If in your little area of the company another major project goes awry, be prepared for extra scrutiny on your program, in a classic case of punishment of the innocent. Your management will be anxious to make sure that whatever went wrong on that other program does not go wrong on yours. You would like to know also what lessons can be learned from their problems.

Often a formal analysis of what went wrong will be commissioned, and findings will be published. Since so much of the success of the program depends on the PM, how that PM worked to understand and resolve problems, manage costs, and ensure customer satisfaction, your senior management will want to measure your competence in those areas. Do not take this personally; it is really quite a natural response. Remember nothing is worse for your senior managers than to be responsible for <u>two</u> major program problems. I have seen some PM's to be considered successful because they perennially *look* worried. While this worried look may make senior managers more comfortable, it does little to inspire confidence in the team. People like to follow successful leaders and the ones who appear to expect success. Confidence in your team is one tremendously important leadership skill—or is it a talent? (It is a skill if you can exhibit it, but a talent if you feel it.) This true confidence in your team is another effect of having a substantial role in its selection. It is much easier to be confident if you have a great team. Building a great team, and aligning their talents to the work is your job. So confidence in your own work is a precursor of your own success!

Program Reviews

One of the most common ways that senior managers assess program performance is by formal program reviews. Very often these are monthly events, but some programs are such that they can be stretched out to a calendar quarter. And at the other end of the spectrum, some program reviews are required weekly if things are not progressing as planned. (Program reviews receive more in depth treatment later in the Section "Monitoring Schedules—Program Reviews.")

Very often the program reviews require the completion of certain prerequisite charts—now that the world has gone PowerPoint crazy, the charts are easier to make, which, ironically, leads to the request for more charts. Figure 7.4 shows you at a program review. (I would suggest a slightly more positive expression on your face and a little more substance in your slides!)

The better work you do on preparing your charts, the higher your confidence will be in the program review. Subconsciously, this demonstration of confidence is what people look for in presentations. It is important not to let your confidence be perceived as arrogance. Such an attitude is even more disquieting for your audience than tenuousness. The risk of appearing arrogant can be reduced by giving credit to your team for the progress the program is making. That is, of course, assuming that the program is making progress.

FIGURE 7.4
Program review.

(From Dollar Photo Club, File #46103811 Klekta Darya.)

Once I inherited a program that was in poor shape. (More on inherited programs in Chapter 14.) As I worked to minimize the losses, the size of the losses became significant enough that reviews were required at higher and higher management levels, until I had to go to headquarters to explain what the problems were and what I was doing about it. (For a brief period I was considered a genius because I found 25% of the financial problem was due to an analytical error: double counting work planned to be done both in house and externally. But you can only find errors that have been made—and the other 75% of the problems were real!) But because I did feel that I was solving more problems than were arising, I did feel somewhat confident that I was doing my best and making a positive difference in the program. I presented my charts to the senior VP, and was pleased that his summary when I was done was: "Thank you Charlie. You have a way of making the bad news sound good." (I was pleased that he didn't say, "Thank you, Charlie. Bring in a more competent PM.")

While it may seem to be a distraction from your day-to-day leadership of the program, the regular analysis of data required by the template charts you need to prepare can actually be helpful. In addition, other members of your team—contracts, finance, quality, etc., will need to give you charts or materials for charts for their areas of responsibility. The requirement for them (and for you) to self-securitize in the preparation of the charts is one of the large benefits of program reviews. It is often beneficial to use these charts in presentations to your team as well. They know the nuts and bolts of the program, but it is good for them to see what is being presented to senior management. It helps them understand your reporting role and may even elicit a little sympathy. Some of your team may think you have an easy job—many of them would find the need to give program reviews daunting—and a little sympathy from your team could help them understand *you* better. For you, after the first several program reviews that you have to give, the nervousness subsides, and you will see it, properly, just as part of your job.

Not that program reviews are without stress. On program review day I would sometimes ask my harried PM colleagues, "What's the best thing about program review day?" Generally, they could not think of anything good about that day. But I always had the answer: "It is the longest time until the next program review day!"

DETECTING TROUBLE AND DETERMINING WHAT TO DO ABOUT IT

Very often when programs are deemed to be in trouble, there is a good reason: they *are* in trouble. Your job is to be constantly vigilant for burgeoning problems and take action to remedy the cause. Depending on the size of the program, your local culture, and your personal skills, you will detect problems through varying combinations of talking to the people and using program metrics. These approaches can be deceiving, and therefore, a healthy mix of quantitative and qualitative watching is indicated.

The key indicator of program progress against cost is the CPI. Described briefly in the Section "Monitoring Progress—Metrics," this is covered in more depth in Chapter 11. But for purposes of this discussion, keeping an eye on your CPI on a monthly basis is essential. How you interpret it is where art and experience come in. As you may remember, a CPI of 1.00 means that you have completed the tasks to date for exactly what was planned for them to cost. Therefore, if you maintain a CPI of 1.00 to the end of the program, that is success. But, of course, that is easier said than done. Remember too, that the CPI indicator is only as good as your plan. CPI's of 1.00 or better often can lull you into a false sense of security.

So how could a CPI of 1.00 be deceptive? If you have ascribed too much value to tasks early in the program, and it takes that amount of effort or less, you will feel positive. But what that means is that tasks later in the program would have to be undervalued—that is to say, it will take more effort than planned to complete those tasks. In design jobs, it is very difficult to predict exactly how long it should take to, for example, design a circuit. Unavailability of parts, inexperienced engineers, and confusing specifications can cause work to take longer than you expected. Keep in mind that the tangible efforts (for example, getting a design complete and working) are more likely to be recognized toward the end of your program. And remember, too, that toward the middle and end of the program, your staffing will be at a peak. The risk that these tangible tasks are undervalued is severe. Early tasks on a program would likely include system engineering tasks, such as identifying and deriving requirements. These are less defined types of tasks and can be considered "done" without the need to, or really the ability to ensure, that they are complete. In fact, when schedule pressure is applied, it is very easy to declare the less-defined tasks as complete.

So the scenario where you have had too many of your precious hours devoted to requirements derivation, which may not be 100% complete or accurate, but are finished for the (too many) hours planned, may make you feel unreasonably confident. In fact, if the system engineering is not thorough and complete, that is another possible cause for the design tasks to overrun—you probably planned that estimate based on 100% accurate system engineering!

All of this points to the need to create an accurate plan—which is what you tried your best to do. Or maybe you did not create the plan—then you would be at the mercy of that person, perhaps your predecessor, to have done it 100% correctly.

Continuing this scenario, on the positive side, you were smart and did not deploy or allocate all the hours for each task, but kept some as a management reserve (MR), so that if things went poorly (or took longer than allocated) you would have some hours/money to solve the problem. Often, that MR is assigned to certain risks, and again that identification and quantification of those risks are subject to your educated guess. If you could indeed foresee all the risks, the PM task would be very easy—but you cannot.

So since quantitative analysis, such as monitoring the CPI, has inherent risks, what can you do to ensure your program is on track? Watching the schedule performance, either manually or using SPI and watching the costs, is, of course, a less quantitative way of assuring progress, but does not make it less valuable. Collecting actual data, by talking to the folks doing the work, is probably the best way to see how you are doing. In the course of talking to the members of your team, you will gather valuable information, and you will be able to remedy issues. For example:

PM: Hey, John, how is the Front End Board design coming?
John: It's going fine.
PM: Good. When is it supposed to go into drafting?
John: I'm not sure—maybe by the end of the month???

> (Gasp—if John doesn't know when he is supposed to be finished, how can he tell how he is doing!)

PM: Let's look at your schedule and see.
John: OK, if I can find it.

> (Gasp—It doesn't seem like schedule completion is at the top of John's worry list!)

So the PM and John look at the schedule together and see that the package is scheduled to go into drafting in one week, not at "the end of the month." This conversation provides an opportunity to alert John to the importance of the schedule and get him help if necessary. If you merely relied on checking your CPI at the end of the month, you would detect that this task is late, but think about it, it would be 3 weeks later! In addition to now having to recover the 3 weeks, you have let John think for those 3 weeks that the schedule was not all that important after all. If it were, he thinks, someone would have come to see him at least by a day after the due date!

So watching the CPI is not a very proactive way to manage your program. It will tell you that things are not going well, but it will tell you that only after things have gone poorly. And, in fact, if your CPI slips, it will take you some time to figure out what exactly caused it. Talking to the people doing the work (or in a large program, one of your lieutenants who is leading that aspect of the work) is far superior to watching the numbers. And it has the powerful and important second-order effect of letting your team know that the schedule matters a lot and that you are watching it closely.

We discussed earlier that knowing just when to intervene when you suspect a problem, and how to intervene constructively, is one of the things that is part of your performance plan. That skill is important to the recovery of the schedule and/or costs. But your skill in this area is moot if you do not first detect that a problem exists. Very often technical people just do not keep the schedule in mind, and merely reminding them of the scheduled events is enough to start the corrective action process on their own. Checking in the following week, and making sure that the schedule is now high (if not foremost) in the engineer's mind, is the way to minimize schedule and, thus, cost problems.

As the program goes on, you will find that your engineers have a range of schedule attentiveness and productivity, and you will be able to concentrate on those folks that are less schedule-conscious. You may hear (and it may be true) that the hours planned are insufficient for the task to be done. But do not be too quick to reach into that management reserve! The best solution to this complaint is <u>we</u> have to find a way to get it done for the hours planned. Notice the emphasis on "we."

If you think of yourself as that engineer, you would consider it unfair to be told to figure out how to do it within the allocated hours. After all, the engineer probably did not quote the hours and feels that applying stress to solve a problem created by others is unfair. This requires more of your leadership art. The engineer needs to understand that no matter how the problem came into being, it has to be solved, and he or she, with your help, is the one who can best solve it. One wrong way to do this is to tell him or her that you: "… want a recovery plan on my desk in the morning!" and then check each day on the tasks that are on that plan. If you do that, you are not only disempowering the engineer, but taking on the responsibility for the task yourself. If you tell him or her exactly what to do (or force the creation of a detailed list) then it is likely that he or she will do exactly what is on that list—and no more. He or she will not be inclined to stay late or come in on Saturday, because it is no longer his or her problem. If you were to sit down with the engineer and develop a strategy together, then the ownership would be in the right place, and he or she would put their heart (and maybe some overtime) into solving the problem.

This performance coaching is actually easy if you just put yourself in the shoes of the engineer with the problem (your shared problem, that is). How would you like to be treated if you were the engineer? Think for a minute and then do just that. You do not get what is needed by dictating. It might feel good for a minute to tell someone exactly what to do and hear "I'll try"—to which you might say "You better do more than try, you better do it!" If you hear "OK" it might just be that's the easiest way to get you to go away!

A poor manager gets lip service; a good manager gets what he or she wants*; a good leader gets what he or she (and the team) need.

WHEN PROBLEMS GET REALLY BAD

By employing the surveillance strategies discussed in the previous section and using more formal EV techniques, you have the tools to detect problems and help resolve them in a timely way. But not all problems are as easily resolvable as the example above. Let us consider a scenario, based on the fact that you may encounter.

Consider a situation where your program team has ten people: two software engineers (SWEs), five EEs, one system engineer (SE) and two mechanical engineers (MEs). You have a contractual commitment to deliver a working subsystem on May 15; it is now May 1, and when the software was loaded into the custom designed hardware, the system crashed.

---

* But his or her "wants" may be wrong.

The customer's team is flying in on May 14 for an early morning start on May 15. You do not like the vision of the customer contingent arriving on the morning of the 15th and you having to tell them that the demonstration is not working. You consider several courses of action:

1. Direct the two EEs who designed the hardware to thoroughly check it out—using mandatory overtime.
2. Direct the SWEs to break their code into 25 bite-sized modules and add them one at a time (actually, you think, they should have been doing this already)—use mandatory overtime.
3. Come up with a recovery plan and track progress against the plan.
4. Call the customer and ask the customer to delay the visit until you get it all working.
5. Tell the whole team there will be 7:00 a.m. meetings to track progress—and in those meetings you will have everyone report.

None of these ideas are very good. Mandatory overtime, especially just for part of the team, is not a good idea. Those who are put under this pressure may feel (and usually justifiably so) that it is not their fault that heroic measures are necessary. Calling the customer is not a good idea—you need to always be building the image of confidence in your team's ability to perform. Missing an important delivery or demonstration is not conducive to that PM imperative. Remember, your customer would have to explain to his or her management that you are late. Hopefully, he or she has been giving positive reports about your progress; now if you are late, an explanation is needed as to why the reports were wrong. Not a great position to build a relationship with your customer into.

Since none of these approaches seem to be very satisfactory or even viable, you will need a more positive approach. My experience is that the best course of action in this situation is to call the team together and give them your nightmare vision of the customer coming in to see a demonstration and our not having anything to show. Lead the discussion on what to do about it. Discuss some of the unsatisfactory ideas listed above to enable them to offer other ideas. The whole team should be included because the more folks you have (within reason), the more likely you are to develop a good solution. Maybe someone suggests that the SE, who is known to be an excellent troubleshooter, should work with the hardware and software folks to find the problem; ideally the engineer volunteers to do so. (People like to use their real talents so this is entirely possible.) Even if you were to come up with this same idea in the privacy of your own office and explain it to the team—it would be your idea not the team's.

Why should someone (the SE) who is clearly not responsible for the problem have to help fix it? Surely, that is not fair! Your response, should that come up, is to agree: "Right, it is not fair." But in a true team, if one of the members has a problem, the whole team has a problem. If the members of the team are as committed as you are to making the demonstration a success, and if you have been able to create a true team environment, you may have to help the team select one of several ideas—none of which would have been popular if you had dictated it. It may well turn out that the team feels overtime is indeed needed to achieve the date. And it may be the case that the "not-guilty" parties think of ways they can help (maybe picking up some of the tasks of the key players to lighten their load in other areas). It is well known that the synergy of group discussion is most likely to solve complex problems—add to that the dedication to the solution that the team's consensus brings to the effort, and even major problems can be overcome. You cannot will success nor can you demand it. You can, however, lead it!

COUNTERVAILING FORCES AND PRIORITIES

As PM, you must constantly balance some powerful forces that affect the quality, costs, customer satisfaction and schedules of your program. Some of your purist engineers may want the design to be so good that it exceeds the required specifications. One of them may want to redesign a working board design because the recurring costs will be significantly less or will "work better." Let's call this "over-quality." The customer may say that a certain gray area of the spec means something more than you and/or your team believes is needed. Do you "give" it to him or her in the interest of customer good will? Your senior management insists that the product go out on time—or actually be pulled in to improve cash flow. Your SWE feels the code she has been expected to reuse is too "clunky" and should be rewritten. The manufacturing team insists that if the drawings are not released by the end of the month, they will miss their dates.

Your job is to sort through all these issues and keep your team engaged, your program on cost, your customer happy and your senior management confident in you. But, it is really difficult. You will need to be reasonable, helpful, and sympathetic—but you must protect the program from unnecessary distractions. As with all problems and decisions, involve the people who are affected by the issue and try to give as much leeway as possible to implement their positive (but possibly expensive ideas).

One of the most powerful forces that could compromise your success is your customer's influence. As with members of the team who are pursuing their own idea of what is right, you will need to be firm at times with the customer as well. Always remember this important principle of contracting—that the responsibility for a clear contract is on the framer of the contract. If you can come up with a reasonable solution to a gray contract area then the requirement is satisfied. The customer does not get to pick the *way* your design meets the requirements, but can just require that it does. If he or she wants a particular implementation and you can do so without undue cost or schedule risk, then it is good to do so—but you should be sure the customer knows that it is a "favor." Not only can you get credit, then, for doing him or her that favor but you help the customer realize that future gray areas may be ones within your power to interpret. If the request is unreasonable, you have no choice but to identify it as "out of scope" and ask if the customer wants you to quote it. Usually, this is enough to make the expensive idea evaporate—but sometimes you may get into a scope argument. Again, working toward a win-win solution by discussion (which of course may be heated at times) is the key to a reasonable solution. The key question to ask in these situations is "why" is this needed or "why" does it have to be done that way? It may be possible to scratch that technical itch by some other simpler approach. Negotiation philosophy is discussed in Chapter 12.

DETECTING AND AVOIDING "SCOPE CREEP"—INTERNAL

This issue of customer preferences resulting in unintended expansion of scope is one of the most dangerous risks to your design program. You must constantly be vigilant. "Scope creep," as it is called, can happen either internally to your project or can be customer driven.

Internal scope creep can occur because your team (and maybe even you) wants to design the very best product that you can. It seems innocuous, does it not, since what could be wrong with "the very best?" Nothing is wrong with it, as long as that is what the customer has stipulated in the contract. Remember, your job is to make money for the company and that means meeting the *minimum* requirements of the contract. It is very unlikely that your engineers will be thinking that way—they will be thinking how good they can make the product, not how "minimal" it can be. Certainly, you do want them to design good products that they and you can be proud of—and hopefully, that is what the customer specified and what you priced.

But there is much gray area in almost every design contract. Ideally, very gray situations are contracted with cost-plus or T&M (time and material) contracts. But in this era of controlling costs, you may find you have a design contract that is fixed price or one that is a hybrid, wherein the financial base is fixed, but there is some cost sharing if the costs exceed the target. Even in cost-plus contracts, costs must be controlled—they are not "blank check" contracts. The customer has every right to expect a certain amount of productive work for the money he or she is spending.

It is therefore important to review the work of your team to verify that the desire to make things better and better does not lead to overruns and delays. You really do not want to find out that your engineers have done too much after too much has been done. So to prevent this, you would be well advised to ensure that they understand that they have been charged with <u>meeting</u> the contract requirements, not exceeding the contract requirements. It is really not that hard to instill this technical frugality. If you explain in your team meetings, and in your one-on-one discussions with your designers, that the customer has bought a certain performance level and that is all we can afford to give, they will understand the concept. You may need to review the concept occasionally, and ask somewhat disheartening questions:

> *Mary:* We finished the range detector feature, and it is 20% more accurate than we expected!
>
> *You:* Hey, that sounds great. What was the specification requirement?
>
> *Mary:* We are way more accurate than that requirement.
>
> *You:* That sounds good, too, but did we spend more than we had to than if we had worked to the minimum requirement?
>
> *Mary:* Well, yes, it was a little more difficult.
>
> *You:* About how many more hours did you spend?
>
> *Mary:* It's hard to say, but it was no more than, roughly, 4 days.
>
> *You:* OK, now we, and that is you and your colleagues, will have to do one of our other tasks in 32 hours less time.

In the end, while Mary is disappointed, she will be more likely to remember the downside of greater-than-required performance, rather than the euphoria at doing more than necessary. To complete the lesson:

You: I'll try to make sure the customer gives us credit for your extra work so we get something for those hours, Mary. But could I ask you a favor?

Mary: What's that?

You: Please keep doing great work, but please check with me if you want to go beyond the contract requirements before you do it. We may have the budget, if we are under-running other parts of the program, or if it is a big improvement, we may want to propose it to the customer and get a change notice. There's no way to get more money after the work is done.

Of course, this lesson that Mary learned did cost you 32 hours. That may not be a major concern on large programs, but, left uncoached, Mary could use this same better-than-required approach again and again, and it could easily cost you more than 32 hours next time. Remember too, that, depending on the type of contract, you may be tightly bound to the scope of the order. Especially on cost-plus contracts, your customer would not want to pay your engineers to "wander in the woods," working on tasks that are not within the scope of the contract. Always respect the customer's money and realize that you are the steward of that money.

Generally, your engineers will be paying little attention to costs and not enough attention to the schedule. They will need reminders all the time to prevent scope creep. If the nature of a task is well understood, and the schedule and hours allocated to the task is accurate, then focusing on completion to the schedule is an important way to control this concern. If an engineer feels like he or she has to spend every hour of his allocation to meet the schedule date, the engineer will be unlikely to add extra work to his or her load. Unfortunately, however, most design tasks are not so exactly estimated, and not every engineer is schedule-conscious or scheduling-able to have that motive as the only reminder to not "guild the lily" by going beyond what is actually required.

So your best defense is constantly reminding the team of the tightness of the budgets and the need to achieve schedules. Then follow up with this theme when you check in with your engineers with that age-old PM question: "How's it going?" If your anti–scope creep message is clear, and if your rapport is good with the engineers, you may get asked about added performance before it is undertaken:

You: Hi, Mary. How's the range detector coming along?

Mary: It's going great, and I think I can get 20% better performance.

You: Really? That could be great. Will it take more of your time to get the added performance?

Mary: Yeah, I guess it will. Should I do it?

You: Well, we need to meet the schedule, and we don't have funding enough for overtime, so I don't think you should.

Mary: What if I do it on my own time?

You: I'm glad to see that you are enthusiastic about it, but if it's a couple of days of your own time, that may be a little too much to give, don't you think?

Mary: Maybe you're right. Don't worry, I'll meet my date and if I can give it a little boost in performance in the time that I can spend on it, then I will. But I do get the message, boss: "Meet the date."

You: Thanks. You know, I'd like to make it better too—but I, and you too, have to constantly be attentive to our costs. It doesn't make the customer happy to have a slightly better product if it is late and if the costs overrun.

Mary: I get it.

It may seem time-consuming to have this conversation, but you can see that these 5 minutes saved you 32 hours, and, more importantly, the discussion reinforced Mary's attention to the hours available for her task and the completion date to which she has been committed.

Scope creep in matrix organizations, can provide another internal risk vector—functional management. The FM should be your ally in controlling costs and maintaining the schedule. The FM can help with your vigilance task by asking "how's it going" and rewarding engineers who exhibit on-time performance and cost control at raise and promotion time. For the most part, you can expect your collaborating functional management to do just that—to be your helper in controlling cost and maintaining schedules. He or she does have another motivation, however, that is not directly aligned with your goals. He or she wants the engineers on your program to grow professionally. Learning new skills is part of that growth. If your engineers are encouraged to learn a new tool in the implementation of your program, it may be good for the engineer, and for the FM (it makes it easier to place that engineer on other programs when the time comes), but it may be terrible for you and your program. Learning costs, and the trial and error that go with learning, are expensive and, thus, a risk to your program's success. Remember that it is common sense that the engineers on your program want to satisfy their "solid line" boss, because that boss typically evaluates their performance. So your vigilance needs to include watching for those "learning experiences." You have to be sure that the FM knows your budgets and schedules are tight; if there is some learning to be done, it should come from the FM's budget, not the program's budget. And there may be ethical issues as well—it might be unfair to the contract to fund the learning of skills expected to be held by the engineer at the time of contract award.

DETECTING AND AVOIDING "SCOPE CREEP"—EXTERNAL

Just as your engineers want the best product possible, so too does your customer. Your program's probability of success is raised by good working relationships with your customer. However, the rapport that you strive for with the customer does carry a risk. You do want your engineers and your customer's technical personnel to work together comfortably. I have seen some cases where the engineers were forbidden to talk to the customer, or at least, not without "adult" supervision. This is of course demeaning and impossible to police—and not at all recommended.

But if your customer and your engineers both want an enhancement that is going to take some more effort, only you can stop it—and it will not be easy. The vigilance that you are using to counter internal scope creep is the same vigilance that you will use to counter external scope creep.

> *Fred:* Jake [the customer] called this morning and asked me a question.
> *You:* Oh?
> *Fred:* Yeah, he wanted me to add a CST [continuous self-test] to the backup power supply.
> *You:* Isn't that out-of-scope?
> *Fred:* Yeah, I guess it is. But it isn't hard to do. It is just a software change— the hardware is already present in the design.
> *You:* How much time will it take?
> *Fred:* I'm thinking a couple of days.

You: Does that include the time necessary to regression test the software and to include modifying the T-spec, both qualification and production? And for each unit, we will have to test that feature as well. It will add to our production test time. Granted, it may only be an hour, but for the 120 units, that's 120 hours of test time.

Fred: Gee, I didn't think of that. That's going to be more like a couple of weeks rather than a couple of days, and if you count the increased test time, it is more like a couple of months! But I already told Jake I could do it.

You: Right, you told Jake you could do it, but you didn't tell him you may do it! Let's call him right now and tell him that the improvement would have much more impact than you first realized. We can tell him that if he really needs the enhancement, he'll have to add it to the requirements. I know it's an awkward situation, but we can't afford the extra costs, and we will have to tell Jake so. Do you want to tell him or should I, or would it be best if we called together?

Fred: I made the trouble; I should take the lead in cleaning it up.

You: Very admirable. Let's call.

So, let us assume that you call the customer, and he agrees that, yes, the new feature is not required by the specification, and he says he will think about whether or not he wants to add it to the requirements. A happy ending and Fred has learned two important lessons: changes are very often much more complex than they seem; and he is not empowered to make contractual commitments to the customer. Because you approached it non-emotionally and because you were understanding about how Fred could think it was a minor change, he will not be overly embarrassed nor will he avoid checking with you in the future. In fact, he will now realize that you have a better grasp of the big picture (that is, after all, what you get paid to have!) and will consult willingly with you in the future.

OK, but what if when you and Fred call Jake, he does not agree the new feature is out of scope? He thinks it is important, and he wants to have it. This is a problem and a challenge, then, for you. You will need to remind him that before this morning, that feature was not even expected, let alone required. You may have to ask him to show you where in the specification it is required. He will not be able to do so, of course. But if Jake is stubborn enough, it could turn into a big problem. You may have to explain to Jake that it is your job to be sure that the contract requirements are met, and that also means not going beyond the contract, even if it is a great idea. Maybe you can avoid the confrontational part of the issue by engaging Jake in a "what can we do about this" discussion. Maybe there is a related change coming up, and this can be worked into that change with little extra effort. Or maybe Jake could commission some new scope—a study of all the CST and its impact on reliability. The results of that study could quantify the value of the new feature and justify a change notice to include it. Again, the way out of a possible confrontation is to consider this as a problem for you and Jake to solve together. Jake and you are probably both problem solvers by education and experience—making it a joint problem to solve together plays upon your common area of strength.

Of course, you can overdo the "no frills" end of the spectrum. Because along with making a profit, your job is to make the customer happy—and a too-close reading of the contract to the minimalist side can really alienate the customer. An alienated customer will not help you when you need it and will not give you more business when the current contract is done. It is, as it seems, a fine line to walk. Protecting you is your and your customer's joint realization that you and he or she will succeed or fail together. If you allow the scope to grow without funding and or schedule adjustment, you will be late and over budget. This is something that you both want desperately to avoid.

This complex balance between what the customer paid for, what the customer wants, what your team wants, what you want, what you can afford, and what the contract says requires a great deal of your skill to manage. The key is, as always, open communication and honest consideration of the viewpoints of others. Through communication, you can control scope and make your customer and your team happy.

SCOPE CREEP—IN SUMMARY

Unplanned, unobserved, and unfunded increases in work can slaughter your schedule and make your program a constant game of catch up, from which you probably cannot recover. Your constant vigilance and attention are necessary to prevent this very powerful force from not meeting the schedule and exceeding the budget. Never do any scope-gray area work with the expectation that you can get paid for it later—most procurements specifically disallow work to proceed that is not agreed upon in advance

FIGURE 7.5
Scope creep.

(From Dollar Photo Club, File #42626194 snyggg.de.)

and cannot be paid for ex-post-facto. Make sure your team understands this principle and make sure that "better" performance of your product is not mistaken for "better" performance of your program. Again, the key is frequent, frank, and direct communication with your team and customer, even (or especially) in areas that they may not want to think about, let alone talk about. Figure 7.5 captures the happy way that your schedule and budget cannot be met, because as you are making the engineers and customers happy, you are slowly moving toward budget and schedule catastrophe.

MONITORING VERSUS CONTROLLING

On large programs you can expect that you will be using EV methodology to monitor costs and achievements. In fact it is mandated on many DoD (Department of Defense) programs. EV is a very valuable tool, but as mentioned earlier, it does have its limitations:

- Its quality is dependent on the quality of your schedule, including estimates of costs and proper task duration.

- It is inherently an after-the-fact monitoring system. You can tell that work has fallen behind schedule (SPI < 1.00) or that it has cost more to accomplish the work than you had planned (CPI < 1.00).
- It is impersonal. You can look at the numbers and have a feeling that you are managing, but, in fact, you are not leading.
- It is an aggregate tool. When the numbers are unfavorable, it is necessary to investigate why. And while you are investigating, the problem is growing. And the aggregating can mask an area of underperformance (and a growing problem) if other areas of the program are doing well.

If you think you are managing your program by analyzing the EV numbers, think of this analogy. The EV numbers are like the speedometer needle in your car. It can tell you how fast you are going, but it is not a control. Your accelerator, brakes, and steering wheel manage the direction and speed of your program. Thinking that you are "managing" by using EV indicators is tantamount to grabbing the speedometer needle in both hands, twisting it upward or downward, and expecting your car to respond. Is the speedometer a valuable *instrument*? Yes, indeed. But do not mistake indicators for controls. You have to do something active to change the speed and direction of your program. And if your speedometer indicates that you are doing 120 mph in a 60 mph zone, it may be too late to avoid the speeding ticket—or the crash.

Thus, you should look at your monthly EV numbers (you may not be glancing at your speedometer often enough), and expect that they will be favorable. That is because you have been actively monitoring progress and costs, watching for scope creep, watching for engineers "stuck" on a technical problem (or a technical nicety). And because you have been doing these things interactively with your team, they will be open with you and will provide early warning of problems that can be avoided, rather than requiring later recovery. So, when you see your monthly EV numbers, you will see that they *confirm* that your program is on target.

It is a good idea in your team meetings (you are having team meetings, aren't you?), to share "the numbers" with the team. The more they understand how the budgets and schedules are watched, the more they will recognize the cost meter and the calendar as part of their design tools.

So often problems are simply attributed to poor communication. The real problem is that the right information may not be getting communicated. Keep your team focused on both technical excellence and schedule/cost performance.

COST CONTROL IN THE TRENCHES

In large companies, not all of the labor that is needed by your program is performed by people exclusively on your program. FMs are a key example. Your EE manager, for example, may have 20 engineers reporting to him or her; maybe only 4 of them are working on your program full-time, and 2 are spending 50% of their time on your program because of temporary needs such as for an Field Programmable Gate Array specialist. At the end of the day that FM (if he or she is a direct charge) will have to account for his or her time that day and charge the correct budget. It is very difficult for him or her to remember exactly how many minutes were spent checking on Mary's progress, how much time the FM spent talking to you versus other PMs that are employing their engineers, how much time the FM spent on training plans, etc. It is therefore not uncommon for an FM to estimate how much time he or she spent on your program. You might expect, in this example, that at the end of the week, there should be no more than the proportion of the FM's direct reports times the 40-hour work week. That is, $(4 + 0.5 + 0.5/20) \times 40 = 10$ hours (Four full-time engineers plus two half-time engineers is one-quarter of his or her work force; thus, in the 40-hour week, you might expect 10 hours.). In fact, you should expect considerably less than the 10 hours, since their time would also include overhead tasks such as personnel planning, training planning, estimating, and reporting. And if those four EEs are senior people who need little supervision, the proportion should even be less. So you check his or her spending on your charge at the end of the week, expecting to see considerably less than 10 hours charged to your budget.

Here is where monitoring can actually turn into control. Suppose you see 15 hours. A call or visit to his or her office to discuss the charges is in order. When you tell the FM why you are visiting and ask what caused the higher-than-expected charges last week, be prepared for a little defensiveness. In fact, there may be a good reason for one anomalous week, but if the 15 hours was accumulated by a simplistic or poorly prepared estimate, your challenge will appear aggressive. Your goal is to not have this discussion hurt your relationship with that manager. What you want to accomplish is calling his or her attention to being more diligent in assessing what charges are appropriate to your program. The mere fact that the FMs know you are watching will achieve this goal. You can say, "Well, I understand that every once in a while the hours will be more than I expect, so I'm not too worried about last week. But if this continues to be over my plan, we could have a problem. Your EE budget includes you so each hour you charge is one hour less for your folks to charge. In fact, your costing rate is higher, so it's more like each hour you charge costs 1.2 hours for your people."

If you do this the right way, the EE manager will be more cautious about charging your job in the future. So in this case you have used an indicator or instrument, similar to a speedometer, to see how fast he or she was charging, and you have applied some control, stepping on the spending brake. Fortunately, through your vigilance, you have caught this charging error in the first week before unnecessary charges were accumulated.

In large organizations and on large programs, this time can be a significant drain on your funds. There may be dozens of people charging to your program—administering your government property, making copies for your engineers, handling customer correspondence, inspecting incoming material, etc. Many of these folks will have the same problem accounting for each 0.1 hours at the end of each day. If yours is one of the largest programs under way, you may get a disproportionate share of their charging, unless you take action. Depending on your accounting system, either you or your FA can look at each person who has charged your job in the previous week. It is well worth a few minutes of your time to scan that list each week and look for anomalies. You may not know all the folks on the list; therefore, it is a chance to get to know them! Such as:

> *You (over the phone):* Hi, Gertrude, this is Charlie, PM on the XYZ program. I saw that you charged the XYZ Program 3.5 hours last Thursday. I'm sorry I don't know you, but I was curious what you did for us?
>
> *Gertrude:* Oh, hi Charlie. Yes, last Thursday I packed a crate of loose parts to return to the vendor because they failed to meet specs. There were 76 parts. Do you think 3.5 hours was too much? They all required special forms and individual bagging.
>
> *You:* No, Gertrude, I'm sure that you charged appropriately. I just wanted to learn what you were doing for us and make sure the charge number was correct. Thanks for your work. Stop up to my office sometime and I can tell you more about the program and what our equipment does for our customer.

So, in this one phone call you found out that a lot of your parts went back to the vendor, and Gertrude is the person who does that work. You may want to find out why so many parts had to go back, a bonus in information for you by making that call. But your purpose—to investigate the charges and find out who Gertrude is and why she was charging your job—was successful. And because Gertrude knows that you are watching your charges and because you treated her with collegial respect, you can be assured that she will be extra attentive to your program and the probability of incorrect charging is thus greatly reduced.

This phone call could have had a different result:

You (over the phone): Hi, Dan, this is Charlie, PM on the XYZ program. I saw that you charged the XYZ Program 3.5 hours last Thursday. I'm sorry I don't know you, but I was curious what you did for us?

Dan: Oh, hi Charlie. Yes, last Thursday I packed a crate of brackets to return to the vendor because they had a plating problem. Did you know about the problem?

You: No, Dan, I didn't. And I'm surprised that we have brackets coming in yet. We are still in preliminary design. Would you please check the charge number again?

Dan: Oops, you were right to check, Charlie. Those brackets were for the YZX program, not yours. The charge numbers are similar, and I must have used the wrong one. I'm sorry about that. I'll go into the system and make that correction right now. Again, sorry.

You: No problem, Dan. Nice talking to you. I'm sure someday you'll be returning my brackets for me. But till then, take it easy.

So you will find little leaks of your charge number if you are vigilant. So be vigilant!

An unkind and often unfair analogy of these small charges is that these chargers, your colleagues, are like fleas, slowly sucking the lifeblood. They may appear to be parasites, like fleas, but you may be failing to realize that they are necessary contributors to your program. Or, they may be true parasites. It's your job to know which!

In fact, while it may seem that when you first look at the myriad of folks charging your program for a half hour here, 0.2 hours there, most of the time you will find the charges are valid and that you do indeed need the services they provided to your program. But, there are certainly cases of people inadvertently using the wrong charge number and using it for a long time before the error is realized. The key is to maintain your charging vigilance. Not to make you paranoid, take a look at Figure 7.6 as a cautionary illustration.

And remember, in your humility, that you could be considered the greatest offender of all! You charge all your time to the program, and yet do not

FIGURE 7.6
Draining your program's lifeblood. (Don't let this happen to you!)

(From Dollar Photo Club, File #21757672 3drenderings.)

(usually) create any deliverables. Organizing, monitoring, and energizing the work is hard work itself, but your time is likely to be pretty expensive. Always look for ways to help your team, even in mundane or subservient roles. Your acceptance of all kinds of work is a great example to your team. One example is going to get dinner for parts of your team who are working late to solve a problem or meet a deadline. If you have 10 folks working one evening, you can keep them working if you bring in dinner. If they go out, not only will they lose an hour of productive work, but they may not have the ambition to come back! Your serving them dinner shows how much you care about the program—and about them. You are going out of your way to serve your team in any way that you can. Check drawings, review calculations, write technical manual chapters, whatever needs to be done—just do it. If you unload some of the less glamorous work on yourself, you could be freeing up technical talent to apply to the program.

MONITORING SCHEDULES—PROGRAM REVIEWS

You should probably accept the fact that try as you might, you will not be able to have your team feel the schedule pressure the same way that you and your senior managers do.

You will no doubt have an opportunity to demonstrate your mastery of schedule and budget at formal program reviews. If there was any doubt that your leadership needs to be focused on getting things done on time and within budget, your program reviews will clarify it. Your first program review after winning a contract is likely to focus on the quality and status of your plans to do the work. Attention will rightly be paid to how well the staffing of your program is going.

The program review content and format will vary from company to company and will also vary with the size of your program. The size of the program will also affect what levels of senior managers attend. Normally you would expect your immediate manager and his or her manager as well. Depending on the size of the program, this may include personnel up to the vice presidential level. You would also expect the FMs who are supporting your program to be on hand. It is a little bit of a show—some of the participants are there to watch.

Ideally, it could be a problem-solving session—since the heads of departments would likely be there so that interface issues could be discussed.

However, it is also a potentially politically charged environment. Remember, the presence of a very senior manager can affect the people at the program review meeting in various ways. Everyone will want (with varying intensity) to look good in front of the senior executives. And even if not wanting to stand out, they would like to at least avoid trouble. Some of your ambitious colleagues may be interested in impressing the executive and may ask self-serving questions.

"Say, Charlie, didn't you notice that the communications board is behind schedule? How could that happen? I thought you said all your boards were on track last month!"

Now that colleague is not out to help you. As tempting as it might be to reply,

"Well, George, it would be on schedule if you hadn't taken Beth off my job and put her on that other project!"

The "audience" will see an unnecessary confrontation and surmise that things are not going too well between you and George.

The best way to handle those comments is to diffuse them as quickly as possible: "Well, George, I guess you're right. But we shouldn't be too concerned because we are making excellent progress overcoming some staffing issues and expect to be back on schedule by the end of the week."

George knows that you are knowledgeable enough about your program to recognize that he could be considered the cause of the problem, and you could have easily embarrassed him. The fact that you refrained from doing so is a credit to your relationship building. People at the meeting who know the story behind the late board will also respect your restraint. If you had "taken a shot," your aggression would be remembered for a long time. The goal should always be to work to get folks to collaborate with you. This may be somewhat odious if you feel that you are getting substandard support from some of your team members, but those issues should be addressed one-on-one, not in front of your mutual boss and not in front of an audience.

So, although all the right people are present in program review and it just seems extra efficient to discuss problems there, it really is not a place for frank conversation. Certainly sometimes, you will need to confront colleagues, such as FMs, whose department's performance is lacking. Rather than in the program review, a much better strategy in dealing with problems is to (politely) confront any of your colleagues in advance of the program review. They will know that the program review is looming, and that you will be performing the review. They will never be more anxious to satisfy your needs than just before the review.

Most of your colleagues will be well intentioned and have the best interest of your program at heart, but your success is very dependent on getting support from all your colleagues. You do not need someone in the program review hoping to see you "in trouble" and maybe even contributing to your trouble.

Remember that your senior managers, one of whom will likely be hosting the program review, will want to be briefed frankly and openly on the status of your program. Do not conceal problems, but do present your plans to solve the problems. It is important that your managers as well as your team remain confident in your ability to pilot the program through the shoals of delays, personnel shortages, belligerent customers, and missing materials. There is a fine line between confidence and arrogance; arrogance will make you multiple enemies. Solve this conundrum by talking about the successes of your team. There should be a whole lot of "we" and very little "I" in your presentation.

The program review intensity will also vary with how well your program is doing. If things are going poorly, you may attract the next higher level of management. You also may get to have bi-weekly or even weekly reviews. In extreme cases, there may be daily reporting—maybe even before normal work hours so as not to dilute your management of the program.

I hope these things never happen to you—but this is not very likely. Because so many things can go wrong on any program, it can easily occur, and you may have this increased scrutiny at some point.

The challenge is to keep these meetings focused on solving current problems, not finding blame as to why you have them. Sometimes the rationale for the "blame game" is to "learn from that error so it will not be repeated." But, often, the "analysis" of the problem is an excuse to punish the person who is responsible, or thought to be responsible, for the problem. Programs are very complex systems, and the probability of accurately and equitably finding the person responsible for a problem is very low. And in the heat of battle (solving problems), everyone should be concentrating on *where do we go from here*, rather than *how did we get here*?

For programs given the scrutiny of "extra" program reviews, it is possible to use these reviews to your advantage. But this is possible only if you continue to think under pressure and look for positive leverage. For example, if your program is having trouble getting the best and most talented people, your newfound friends in high places have the power to get the people you need. If you can effectively make them part of your team, rather than your panel of judges, you can actually turn your program around quickly.

Having your senior managers directly involved in your recovery will greatly help. They will realize the size of the challenge, and they will share in the success, as things get better through their help. Remember, their motivation is identical to yours—get the job done, done right, and return to schedule. It is unlikely that their attention will increase your personal motivation, since you are likely to be highly driven to succeed anyway. But it is entirely possible that some of your team will be more motivated with such senior visibility. Ideally, you built your team with self-actualized folks, who want to succeed for the good of the program and for their colleagues—including you. But, nothing is perfect. You probably had to make some compromises in whom you took on the program. Some of those people may in fact be motivated by wanting to impress the managers who have suddenly taken an interest in your program. Some of them may be afraid of reprisals for failure—lack of raises or bonuses, or maybe even termination if things get bad enough. Fear is a powerful motivator—but its side effect, anxiety, is not good for anyone.

This is especially true in times of possible layoffs. Everyone is very driven to succeed on the programs that they are on since the people on vital, successful programs are more likely to survive a layoff. But if the team is sufficiently worried, then they will be talking about what rumors they may have heard, or they may talk about other companies in the area might be hiring. When people are anxious and under stress—they do not do their best work. It may be hard to concentrate and collaborate.

I had only been a manager for a few years when work got slow, and layoffs were beginning to be of concern to the community. You could tell that the engineers were distracted. Even though the folks on my program were very talented (and importantly, they were recognized for their talent), and even though our program was going well, people were still scared. I began to notice that when I went to someone's cubicle and asked them to come to my office, they might flush, pale, or even jump. And after this happened a few times, I realized that I had been mistaken for the Grim Reaper. I was not calling them to my office to lay them off; I was just calling them to my office to work with them on some aspect of the program. I soon learned that instead of, "Dale, would you please step into my office?" it was more polite (and really more humane) to say, "Dale, I want to talk to you about the EMI (electromagnetic interference) issue. Could you step into my office to discuss where we are on this?"

So fear is a distraction from intellectual pursuits. But seeing people appear to be working hard might make you and or your senior managers feel better. It is the old transference of stress foible—if the workers are worried, then they will work harder, and I will not have to worry. That is quite a false comfort. If your team were lazy and not working hard, then maybe a small dose of fear might help—but that is almost never the case. Engineers, as a group, are generally the opposite of lazy. They went to school and chose to take a challenging curriculum—they are almost certainly made of the "right stuff." Do not try to scare them into better performance—it just does not work. Encourage and help them.

LEADERSHIP AND CARING

The best leaders honestly care about the people they are leading. Managers do not really have to care, but remember that effective program management is where leadership and management coexist. Caring about the emotional and physical health of your people, and letting your concern be known, is one of the most powerful leadership models around.

At times, this can lead to some personal conflict or at least some personal confusion for you. Suppose you are faced with a deadline, and you have three engineers who need to work together to solve a problem, for example, before the customer comes next Thursday for a meeting. You are fortunate that you have a very smart and dedicated team, but like many technical endeavors, it is impossible to tell how long it will take to solve the problem. You know and the team knows that no one will be comfortable about the visit until the device is working. It is already later than you and they would like. It would be much preferred to have the unit working and test it thoroughly days before the customer visits. But you do not know when the problem will be solved.

Your team is so dedicated that they have decided to work through the night and over the weekend on the problem. You are justifiably proud, and you are smart enough to buy them dinner so they can keep on working. But it is Fourth of July weekend, and one of the engineers has an out-of-town family commitment. Old-school management would say, "Work comes first!" and with enough power and maybe threats that could keep the team working.

But by dictating when and for how long the team works, that old-school mentality actually removes ownership and motivation from your team.

A good leader would look for a way to both get the work done and respect the out-of-work commitments of his or her team. You do not have to be very smart to do this—really, all it takes is trust in the dedication, drive, and ingenuity of the team. Those three people know exactly the course of action to take and know all the "Plan Bs" that will allow them to work around problems. If you have set the tone of the program correctly, and if you trust your team, it is very likely they will work out a plan so that their colleague can keep his or her family commitment and still make good progress on the problem. But this can only happen if they trust *you* to trust them. If they feel they cannot approach you with a Plan B without making you angry or resentful, they just will not develop a Plan B. Remember, if you treat your team with understanding you can expect that respect to be reciprocated.

In fact, during stressful periods on a program, some team members will burn out and become ineffective because of exhaustion or stress. Your job is to *strategically* draw the best work from your team—and that may involve some slightly uncomfortable *tactical* choices.

This story is one of the best examples of leadership that I have witnessed. I attended a program review as a "spectator" on another PM's program. Things were not going well to say the least. The design was late, the customer was angry, the budget was overspent. And as I mentioned above, the poor state of the program had attracted the attention of senior management. In fact, the newly appointed VP, whom none of us knew, had come to town to attend the review. He listened quietly as the PowerPoint slides were projected and explained. It really was bad news. Through his silence we could not tell if he was being respectful or was, in fact, seething. So when the PM, Barbara, finished, the VP said: "I have a couple of questions, Barbara." Well, Barbara expected something like, "How the heck did you let it get like this?" But the VP said—"I understand that the team has been working hard on this program and has been putting in a lot of extended hours and weekends. Are they burning out, Barbara?" Wow–concern for the team! Barbara said that yes, they were all pretty tired, but they all expected to turn the corner soon and that they felt that the end was in sight. VP: "That's good, and how about you Barbara, are you doing OK? " That the VP, even in times of stress, showed concern for the people was effective leadership, and those two questions did more for the ultimate success of the program than any retribution-oriented inquiry could ever produce.

Remember, exhausted engineers just cannot do as much work as those with energy!

PROGRAM CHANGES AND CONTINUITY

As we have discussed, there are all types of programs and all types of contracting vehicles to manage them. One of the most challenging is a long-term program that begins as a study, progresses into conceptual design, moves on to detailed design, then preproduction fabrication, and preproduction (qualification testing), production, and finally, field support. While often these phases are contracted for by different types of arrangements (for instance, T&M during the conceptual design and fixed price during the production phase), it is extremely possible that there may be a single PM for the entire life cycle of the program. Customers and your management may feel more comfortable, assuming you are doing a good job, keeping you in place for the entire program. This could be several years in duration—perhaps up to a decade.

While your personal primary competency may be in the design and development phases, other considerations, such as the customer's comfort with you, may result in this type of very long-term assignment. There is a comfort for you in the continuity also, but your continued growth and challenge is important to you and also, because it affects your effectiveness, to your company as well. You may need to leave the program and bid on something new, or you may be happy to stay with your team and derive satisfaction and new challenges as the program progresses through its various phases.

But what about your team? They are faced with the same choices, too, as the program progresses. The first part of your program will likely be system-engineering heavy as the trade studies and requirements definition phases of the program are performed. As you move into conceptual design, however, you will need a different mix of engineers—perhaps more senior EEs, MEs, and SWEs—to develop the concepts and system architecture. As you move into detailed design, you will probably need design engineers, and these engineers may include some junior people, working on details under the tutelage of the senior technical staff that you built in the previous phase. As the detailed design gets done, you will need engineers who can follow the early phases of fabrication and production, modifying drawings as discoveries are made in the first build in your shop. When the preproduction units are built, you will need engineers to test the equipment in the specified challenging environments. You will need staff to write the qualification report. People will have to write design reports and technical manuals. You will need engineers to design test tooling, and to follow production, solving parts and fit issues, finding and fixing small software bugs. You will need members of your team to support installation and customer training. Finally, you may need staffing for field support, solving technical problems, or installing field kits. It is unlikely that any one engineer has the skills or the desire to stay with the program through its whole life cycle.

Your program, however, benefits by maximum stability in your team, at least to the extent that the engineers' skills and talents apply. An engineer who has been on your program for a period of time will have learned and understood the requirements, will have formed bonds with his or her teammates and may even have developed a rapport with the customer. Clearly, it costs your program money to bring a new person in and bring him or her up to speed. But continuity must take a backseat to competency. A pure SE should not be asked to design circuits, but there may well be need for an SE throughout the first several phases of a program, as customer expectations, requirements interpretations, and team learning are still a full-time job—at least for one person. And some of your design engineers may decide that they can write plans and/or run qualification testing. In short, some of your team "specialists" may break through their specialties and remain with your program for multiple phases.

You may come to rely heavily on some engineers. Depending on your own technical strength, you may need some of these folks to provide technical leadership and/or to interface a very technical customer.

While you may *want* some engineers to stay on the program, you must appreciate that you do not have control of their assignments. Even if you have sufficient rapport with the FMs, and even if you have enough organizational "clout" to hold the engineers that you want on your program, you should not do it unilaterally. If a key team player feels that his or her career is compromised by remaining on your program, you must accept that, and even help him or her find a good next assignment. No one works well when they feel captive. You are considerably better off strategically if you allow, and even help, a valuable team member leave as soon as he or she wants. If you help someone find a good next assignment, (and if you make it clear how valuable you feel he or she has been to the program), you can expect that said individual will be willing to come back to you on your next program. Your "control" over your engineers is not through direction (management) but through attraction (leadership). The best leaders will naturally attract the best talent, and the best team will perform remarkably well.

Keep in touch with the folks that move on to other programs. Very often you will have formed very cordial and professional relationships with them. Knowing them and their skills, you will expand your leverage, and you will know whom to call upon if you have trouble. "Networking" is sometimes a pejorative word, but when you have a relationship with a person based on mutual respect, preserving contacts is good for both you and your erstwhile teammate.

Remember to celebrate the engineers that move on to other projects or programs. You can thank them publicly at a team meeting, you can get a cake or a pizza to celebrate, or you can simply acknowledge them privately in an e-mail. It is great if you can recognize them for some particular important work, or if you can recount a humorous incident involving the departing colleague.

Make sure also to be attentive at the other end of the cycle. You will definitely want to be sure that new staff added to your team is introduced around and made to feel welcome. Just think about how you have felt when you were added into a new social situation and how uncomfortable it may have been. Perhaps a peer or a leader recognized your need and helped you acclimate to your new role and colleagues. As the team's leader, you have a particular obligation to make sure that your team works together, and this act of consideration makes your team more effective, as well as showing you to be a caring leader. Caring leaders are the best leaders.

One of the more surprising things that I have seen is how remarkably easy it is to lead the team over these transitions. In general, the right people and even the right number of people opt in and out of your program at each transition. This miracle is lubricated (to mix a metaphor) by flexibility, both on your colleagues' part and on your part. The basis for the miracle of self-selection is that trait discussed earlier: people like to work in environments where their talents lie. Engineers with poor writing talents will naturally shy away from writing design reports and technical manuals. Thus, when it comes to the writing-intensive part of the program, if allowed, the natural selection process of folks choosing work that aligns with their talents will help get the right people on the right tasks. The old-school thinking was that team members will do as they are told, and the less glamorous tasks (as some feel technical writing is) must be shared by everyone. If you are able to establish a good team environment, there will be a natural sharing of the less agreeable tasks. And more importantly, given enough freedom, people will gravitate to tasks that use their talents. If they are good digital designers, and that is their only talent (viz., they are not good writers) you would expect them to be looking for more digital design work. And someone who is both a good test engineer and a good writer may offer to write technical manual chapters after the test specification is finished.

As with almost all leadership work, the key is to listen to the individuals you are leading and to care about what is good for them. Believe it or not, what is good for them is good for you. If you lose a key player, your job is to resolve that issue. The stronger your team the easier it will be, but remember, dealing with program issues is your job. As people come and go on your program, it is up to you to maintain progress and efficiency. It might seem easier if you "strong-arm" someone into staying on to the next phase, asking him or her to work outside their talent range, but that is just a short reprieve. Good people make poor captives, and you will likely face that transition in the near future anyway. It is much better if that transition occurs cooperatively rather than antagonistically.

MANAGING EXTERNAL CHANGES

We have been talking about dealing effectively with changes that occur in the natural course of a program. But more challenging is dealing with external surprises. These surprises are not of your making, but you must deal with them, nonetheless. Here are some examples of these external (to your program) changes:

- Customer funding abruptly changes—you will need to adjust staff quickly and perhaps not at a natural break point.
- A key member of your team is transferred.
- A new general manager or VP is appointed and wants things done in his or her own way.
- Your division is relocated to another company-owned building on the other side of town.
- Your customer changes scope, and a new plan must be developed, quoted, and executed without adverse effects on the program.

Clearly, these are events that cannot be planned for in advance. How you lead your team through these challenges is a true test of your leadership and maybe even of your character.

I once had the privilege of seeing good leadership through crisis at another division of the company. A surprise funding cut on a big program caused the displacement of dozens of engineers and a large potential for layoffs.

Since our division was doing well and looking for talent, I went with a colleague to interview candidates. The program leader was accommodating and solicitous regarding our getting to meet the most and best candidates. We spent all day working with him in attempting to find the engineers with the right talents for our needs. At the very end of the day, the leader (who was very senior in rank) said: "And would you have a position for me?" Clearly he had put himself last. Not only was it more important to him to have us spend time evaluating his team, but by asking last, he ensured that he did not take a slot that could be filled by someone else. To me, this is the essence of leadership—putting your program and your team, and even the folks on your team's individual best interests, ahead of your own.

CELEBRATING VICTORIES—CONFRONTING DEFEATS

As the program goes along, there will be many in-process successes. A deliverable report is prepared on time (or a little early—allow for that final reading discussed above!). A troublesome circuit problem is solved. A customer visit goes particularly well. You reach a milestone on time—the first submittal to drafting is on time and of high quality. These little victories are not only progress in themselves but they are also symbols of your progress, and thus they are portents of your program's success. Celebrating these successes continues to invigorate the team. It does not have to be elaborate—just move your weekly team meeting to lunchtime one day and order in pizza. Those "free lunches" are one of the best investments you can make. Not only does it prevent your team from having to go out to get (and pay) for their lunch, but it also breeds team camaraderie in a natural way. No need to play contrived "games" or other team-building events— eating with your teammates is really special. It probably harkens back to our ancestors who would take time from the hunt for lunch together.

Greater successes may call for greater celebrations. Here you are likely to be constrained by your company's rules and or culture. But remember that as the PM you are also the chief publicist for the program. It is absolutely part of your leadership role to get your teammates recognized for their accomplishments.

On one of my programs, we had just completed a strenuous and tense preproduction test program. We had to work double shifts to maintain the schedule and to minimize outside facility costs. Team members had to travel for days at a time, missing valuable time with their families. Good teamwork and fellowship surfaced, with engineers who were between family events, filling in for those whose children were having birthdays or important little league games. Nonetheless it was grueling. Furthermore, in preproduction testing, seeing your designs challenged by extremes of temperature and humidity is really stressful! Of course, not everything went smoothly— test equipment went out of calibration, shock table availability windows were exceeded and tests were delayed, and test tool failures needed to be resolved. Furthermore, this tense period went on for 2 months. We had plenty of customer visits, and their tension just magnified our own. And then... we were finished. Whew! All the tests actually did pass and we could go home.

This was an event worth more than a pizza lunch. We were able (especially since senior management was sharing the tension of completing the testing) to obtain some budget money to take the team out to a nice dinner (Ironic in that there had already been many meals "out" between tests, or after late shifts.) Of course, the PM had to make a brief speech and opened the floor to others who wanted to say something. One touch that added to the celebration was a little "award ceremony." We got small plastic trophies and made up plaques for them that recognized the team members for their particular contributions or even their quirks. We called each member of the team to the front of the room to present the trophies and read the inscription. One of the engineers (an SE) was the "go-to" guy and was recognized as "Our Expert." One of the folks led the EMI tests, one of which included building a huge coil of cable and positioning our equipment in the center of the coil. From a distance, and in the photographs that documented the testing, the coil looked as if it was a hot tub. So Tom's trophy said "Keeper of the Hot Tub." Not only did these semi-silly awards bring lots of laughter at the dinner, they also were a tangible symbol of the camaraderie of the team. You could tell it was successful, because for years you would see these little trophies on display in the engineers' cubicles. They made for nice little mementos and conversation pieces. Recognition and celebration are really much better if they are warm and a little different. It makes them memorable and shows that you took the time to prepare for the event, and more importantly, that you were able to recognize the contributions and even give them a positive twist while being funny! That is a challenge. But remember, on your team, there is someone who would be great at coming up with those ideas. As a good leader, you know who it is and let him or her use their talent to help you. Recognizing talent and putting it to good use is a hallmark of a good leader.

It is likely that you as a PM are an engineer by training. And we engineers are stereotyped to be conservative, or shy or withdrawn. And there may just be something to that. It may not be easy for you to discharge your "social" responsibilities. But you are being paid to be an effective leader, so you should not shy away from the uncomfortable.

> Q: How do you tell an extroverted engineer?
> A: He or she looks at *your* shoes when he or she talks to you!

As uncomfortable as it may be for you to preside over celebrations of successes, it is far more difficult to confront failures or disappointments of the team. But the ability to deal with all news, positive and negative, in an open, forthright manner is vital to the success of your program. Suppose the customer has a negative reaction to a document your team submitted. The lead engineer on that task will no doubt feel terrible (or angry). His or her sub-team will also be disappointed. You may have a tendency to minimize the damage by not discussing it in your team meeting—after all, it is uncomfortable for everyone. But that is a mistake. In general, if you avoid something because it is uncomfortable for you, the avoidance may actually be a minor act of cowardice.

It would not be a bad idea to talk to the sub-team or the sub-team leader about your plan to discuss the negative feedback in the next team meeting. It may not be comfortable for them, either. But the discussion can become a learning experience for the rest of the team, and for the sub-team that is going to submit the next report. Why did the customer not like the report? Were there some valid points in his or her complaint? Were there some invalid ones? Was there a miscommunication of expectations? Perhaps we should push back on the customer. Some of the discussion may lead to action items for you if it is decided to push back on some issues. Or maybe the failure to clarify mutual expectations is actually your fault. Surprise! You are not perfect either. Your ability to take your share of the blame (and maybe a little more) cements your leadership image and position. A little well-placed humility and a willingness to face your errors and shortcomings is part of the profile of a "thinking-person's leader." Ironically, you making a mistake can be very valuable to your team. Handling it well is the most effective way to encourage your team to promptly own their errors, and own the mitigation. The best leaders are flawed leaders who are constantly interested in learning from their peers, their own leaders, and in this case, from their mistakes.

DEALING WITH INDIVIDUAL PERFORMANCE PROBLEMS

As we discussed earlier, sometimes the team will fail to perform to your satisfaction—a date will be missed, a design flaw will cause a significant failure, or a poor quality report is sent to the customer. Team failures are one thing, and need to be addressed with the team. But what if the problem is traceable to one person failing to perform adequately? Or what if someone on the team is failing to do his or her share of the work? It may be difficult for you to detect a particular inadequacy; there may be dozens of people on your program, and your contact with some of them may be rare. So if it is unlikely that you will have first-hand knowledge of a significant flaw, your information may come in another way—one of your team members may come to tell you about Jim's performance.

This is a difficult situation. You will have to determine if the person who has come to you (for example, Mary) has pure intentions and a circumspect view of the situation. Perhaps Mary has a personal dislike for Jim and wants to see him reprimanded. Or perhaps Mary's negative view of Jim colors her observations.

Like any "news" that a member of your team brings to you, you must welcome the information. Even if at first it seems like someone is treating a peer unfavorably, withhold your judgment. Also, at the other extreme, it could very well be that there is a problem with Jim's performance. Your best course of action is to thank Mary for bringing you the information, affirm the fact that the entire team needs to work to high standards and that you care about everyone's performance. Avoid agreeing or disagreeing. Avoid either minimizing the concern or over-reacting to it. It is probably best to make this conversation relatively brief, lest it becomes more negative toward Jim. Remember, your goal is to encourage people to do their best and to work harmoniously with their colleagues. You can do this best when you have all the information you can get. Sometimes the signal may be buried in noise, but there may be a signal there. You are much better receiving all information and sorting through its validity off line. Your investigation will be more effective if you ask the person reporting the problem to you to be specific.

For example, it sure would have been easier for you if Mary never came to you in the first place, would it not? You could have referred Mary to

Jim's FM: "Hey, why are you telling me? You know Jim reports to Mark!" Of course, you would not do that because, if you did, Mary would assume

- You do not have the courage to deal with the problem.
- You do not care enough about the problem or the program to act upon it.
- You do not value Mary and/or her information.

Yes, it would be easier if you ignored the report or if you sent Mary off to Jim's FM—but this example is your responsibility—you do not get paid for easy. You want all information, regardless of whether it is suspect, positive, and negative. Once you have the information, you can make a decision on what actions, if any, you should take. But if you do not get the information, you clearly cannot take any action. This even pertains to complaining— even if the employee is complaining to you about health care costs, you allowing him or her to vent may avoid brooding. Brooding engineers do not produce as much work as non-brooding engineers!

Depending on the type of negative information, you may need to involve Human Resources or Security. If the allegation is about sexual harassment or improper conduct, report the issue immediately to human resources, being clear that you have not investigated the allegation and have heard of the issue second hand.

But performance issues <u>may</u> best be handled by you. In theory, as a PM, you are supposed to be relieved of these issues by the FM of the alleged miscreant. (Always remember *the alleged*). Your choice on turning over the alleged problem to the FM depends on one primary factor: who, you or the FM, would do a better job of investigating the allegation and taking any necessary action. Think of yourself and the FM as part of the leadership team. Just as you would want your engineers to parcel out work to the most capable member of the technical team, you and the functional manager can work together to decide who should handle it and how. Either you or the FM may have a long-term and/or good working relationship with Jim. For example, if you have worked with Jim closely for 5 years, you may be the better one to investigate the problem. Or the FM may be very skilled in this kind of issue and then he or she do the investigation, and if necessary, the help to solve the problem so it does not reoccur.

Let us say for the sake of this discussion that you and the FM decide you should investigate. A poor way to investigate is to ask around, checking with other members of the team. That puts those folks in an awkward position and may improperly injure Jim's reputation. The best way to investigate Jim's performance is by talking to Jim. And there is a right way and a wrong way to do it. The worst way would be to say, "Mary tells me that you messed up the design of the preamp board; is that true?" If there was animosity between Jim and Mary, this type of question would exacerbate it; if there was not, this question could create it.

Another risky way to make a determination of Jim's performance is by surreptitious observation of Jim at work. If you walk by Jim's cubicle and he is on a personal phone call, you may make the rash assumption that he makes too many personal phone calls. It is probably impossible or at least impractical to get a fair reading of someone's performance this way. Beware of isolated observations!

This is a story about how it feels from the other side. I was a young engineer, and at the time the program I was on was in a bit of a crisis. I was working extended hours and was fairly stressed out. One day, I asked one of the other engineers a technical question, and he wrote it down and made a paper airplane of the answer and sent it to me. I told him, "That's not how you make a paper airplane!" and I made one (actually, a much better one) and threw it back to him. Well, it flew well, but not straight, and it veered right into my boss's open office door. (See Figure 7.7, which shows some communication ideas that were not the best.) My boss knew or suspected that I was working hard and did not say anything when I went in to retrieve it. However, a week later, he called me into his office to tell me it was unprofessional to be making paper airplanes and throwing them around the office. I told him the circumstances, and he told me why he had spoken to me: one of his peer managers said that he saw one of his engineers (me) who apparently had so little to do that he could make paper airplanes and throw them around the office. So although I could understand that I had put my boss in an uncomfortable position, I didn't think that this one observation should be taken to be that I had not worked hard! Hours and hours of extra work were ignored, and 30 seconds of airplane making were magnified. While we might question the motives of the other manager, the lesson that I learned from this event is that if you are going to make judgments about others, you had better be more careful in your observations. It is really a signal-processing problem: what sampling rate do you need to measure performance. Personally, I think it is so frequent a rate as to make it impractical. Find other ways to measure performance!

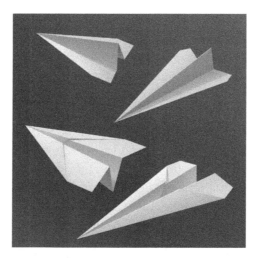

FIGURE 7.7
Unapproved intra-office memo forms.

(From Dollar Photo Club, File #6180156 Nanshiro_son.)

Thus in the case of Jim's performance investigation, it is much better to say, "How is the design of the preamp board coming, Jim?" Depending on the answer, you may need a few follow-up questions to determine if there is a performance problem. And you may need more than one session. Here you can use date leverage to measure how it is going: "When is it scheduled to go to drafting?" Hopefully, Jim knows and can tell you he is on schedule. If not, there is reason to increase your surveillance. Another great question to draw out performance issues is: "Is there anything holding you up or giving you trouble?" Depending on what information you got from Mary, you may be able to approach the area of concern. If you keep Jim relaxed and comfortable, you have a much better chance of detecting and diagnosing a problem. Remember too, that almost all engineers are ethical about charging their time and that almost all engineers are not lazy. If Jim's performance up until now has been good, it is unlikely that he changed; other than because of a problem, either personal or work. You do not need to get into his personal life, but your conversation(s) with him should make him feel supported and valued, which may be extremely important if he has active personal issues.

On the other hand, in the unlikely event he has not been doing his assigned work, he will now know he is being watched. Always make it easy for your team to talk to you. The more they talk to you, the more you know, and the more effective you will be.

If it turns out that you decide that Jim's FM is the better person to talk to him, it may be good (and maybe just a little invasive and pedagogical) to ask him or her how they intend to approach the potential problem. If they have little or no plan or if they have a confrontational or gossip-mongering plan, you have the obligation to comment. A disenchanted Jim hurts *your* program, and you cannot afford for that to happen.

After your investigation (either by yourself or with the FM), get back to Mary with another thank you and let her know that you looked into it. Depending on what you found out, you may or may not want to share it with Mary—it is important to respect Jim's privacy and important not to share any personal negatives that you found out along the way. If it is something non-personal, and Mary's comments have led to an improvement, then it would be good to share the details. This would be something like—"Jim didn't know that the standard parts list also covered capacitors. He spent a lot of time looking for just the right part. I wonder if all the other EEs know what is included in the standard parts list? I'll bring it up at our next team meeting—please remind me, Mary, if I forget."

DIAGNOSING AND RESOLVING PROBLEMS

As we begin this section, it is critical to understand the difference between instrumentation and control. You use your instruments to detect problems, but you cannot solve problems using the instruments. Monitoring your CPI can alert you to a problem—but you are going to have to determine why it is eroding. And then you have to take action to resolve the problem. This is the "control," and the most effective approach is to use the relationships that you have built with your team.

The problems that you face as a PM are both numerous and varied. Most people who enjoy program management are people who enjoy solving problems. As an engineer, you might get to solve a single class of problems, but PMs have problems in many dimensions at once. Some of these problems are technical, and have to do with specification conformance and the customer's perception of the specification. Other problems are parts related—how can we get this component faster to meet our schedule, or can we find an alternative? What should we do about an error we found in a delivery date in the contract? Is the number of printed circuit boards we are seeing returned for repair beyond normal or within acceptable limits?

There can actually be an advantage to complex problems. The fact that they are complex means that you have many parameters with which to work. If you have good relations internally and externally, many more options are available to you to help solve these complex problems. And remember, you do not have to solve these problems by yourself. You have many smart people working on your program, and in your organization, and describing your problems to them will allow them to use their creativity to help you. Much like you, they like to solve problems and help people as well. Never be too proud to ask for help or advice. Pay no attention to the person's rank or status. Senior managers and test technicians also want to contribute to the success of your program. It is good, however, to ask the right people for help. Hint: Ask people who are helpful! And avoid those who are not! (Another example of how helpful people get information, and are able to help: The folks on your team think like you do. They will ask you for help when they expect you are going to be helpful, and they would work around you if they think you will not be helpful!)

What all these situations have in common is that the solution lies in working with other people to understand and then to resolve the problem. Some of these people are not part of your program—for example, the pack and ship engineer. So you can rule out your "power" as the tool for the solution. Managers use power—leaders use influence. It is great you have leadership talent! Your relationship with the people whose help you need is critical to your success. And for people whom you do not know, your *approach* is critical. Suppose there is a new packaging engineer in your building, and you must meet the schedule to get something to ship. Here are a couple of approaches for you to consider. Which is more likely to work?

1. Phone call: This is the PM for the XYZ program calling. I'm sending down three cabinets that need to go out tomorrow. It's very important that they do. Let me know when they ship.
2. Visit: Hi, I'm John Smith, are you our new packaging engineer? Did you just transfer in? How are things going, do you like it here? I hope you're not too busy. I have a critical shipment and I'm going to have to ask for your help in getting it out tomorrow.

I think most people respond to approach 2. Why would anyone use approach 1? One answer is that they are too busy for pleasantries. If you measured how much longer it took for approach 2, you would realize how specious that frequently-used excuse is. One real reason someone might use approach 1 is that they feel they are too important to waste any time on someone who is not a manager. Or they might actually not think of the packaging engineer as a person, but as a function. If you do not see the people you work with as people, but as resources or minions, you are not a leader. Certainly, as the PM you can "throw your weight around." And at the time, you may get apparent subservience. But your arrogance has prevented you from seeing the next step. So if you used approach 1, after you have finished talking with the new packaging engineer, she had reprioritized her stack of shippers, and yours probably was now on the bottom of the pile! At 5:00 p.m. the next day (the day you want your three cabinets to ship), you call the packaging engineer and say, "Why didn't you call me when my cabinets shipped???" You will probably hear that there were other priorities, and your cabinets will ship "soon." Intimidation only appears to work—it really does not work. Consideration, understanding, and humanity always work.

CELEBRATING THE SUCCESS AT THE END OF THE PROGRAM

It is nice to envision the completion of the program—all the equipment has shipped on time, all the reports are approved, the customer may have come down to thank the team. But one warning: Programs do not typically end with a single, visible event. Even after your last shipment, the equipment still has to be installed on the end platform and integrated into the other systems on the ship/plane/vehicle/plant. Members of your team are likely to be performing this work, possibly under another contract, but they are still considered by everyone to be on your program. Thus, it seems like the wrong time to celebrate the completion of a program when the last shipment is made. Maybe you are still writing technical manuals or cleaning up reports. So if you are thinking that you should save all your celebrating for the end, that may be a bad idea. Furthermore, if it is truly the end of the program, it is right to thank the people who made it successful, but the need to motivate them on your program is over.

Hence it may be better to celebrate major milestones on the program, since there in addition to thanking the people you also get to re-energize them, doubling the value of the celebration! One of the best motivators is appropriate recognition for a job well done—for the team that comes at these times of celebration.

SUMMARY

This chapter contains some of the basic concepts and strategies for running a complex program. A design and development program environment was exemplified since that is where the largest risks are, and thus the most PM skills are needed. Important and detailed concepts, such as claim avoidance and claim management (Chapter 8) and leadership models (Chapter 9), are covered in more detail in separate chapters so as not to interrupt the flow of this chapter's discussion.

In this chapter we examined some basic concepts of how different styles of leadership affect the team's attitude and motivation. We presented ideas about how you can monitor progress, and how you can be aided *and* deceived by metrics and EV concepts. We also examined the difference between instrumentation (knowing there is a problem) and control (taking action to fix the problem). Some ways to collect information are superior to others, in that they lead to a quicker and more complete diagnosis of problems. Quality and your constant attention toward it is of the highest importance, and your role as the leader of the program and your attitude and focus on quality is at the heart of a successful program and at the heart of repeat business for your company. We also differentiated the concept of quality from the concept of perception of quality; you may have a quality product, but that is unimportant if it is not perceived as high quality by your customer. And, speaking of the customer, we discussed how the customer can influence your program and your team, and how you could use the customer and his or her visits to help motivate your team. You were warned of some common performance traps and coached on how to keep your engineers moving in a forward direction, allowing only a reasonable time to celebrate and make improvements along the way.

Senior managers (your manager and the engineers' manager are probably not the same person) are part of your management/leadership system and they too (obviously) have a stake in and an influence on the success of your program. You have an obligation to them to keep them informed of problems and to seek help as appropriate and discuss alternatives as needed. (Just as you would want your engineers to ask for help when they were in trouble, you must have the humility—and ironically, the self-confidence to do so yourself.) As the PM, you are constantly balancing and making trade-offs, and the ability to do this in a positive way is shown to be the path to success. The age-old monster of scope creep, which often masquerades as "improvement" or "quality" is addressed. We discussed its sources, internal and external. And we discussed its insidious nature. That happy little snail illustration (Figure 7.5) is a terrific symbol of how it can get you!

In large companies and on large projects, your program will need help from a surprising number of people that will "direct charge" your program. These services are often overlooked and are frequently episodic. Sometimes these helpers are seen as negative stakeholders—and sometimes they are. But vigilance as to who is charging your program is the key to your immunity from the disease of death by 1000 cuts—a few hours here and there can add up to a serious drain on your program's budget; a few constructive phone calls can assist your helpers make sure they are charging appropriately and not out of habit.

We discussed program reviews. Program reviews are opportunities for help, but they are also opportunities for misunderstandings. You can make friends or foes in what you say in these meetings, because they are typically in front of senior managers, and thus raise the level of anxiety in your colleagues, who may fear that a poor impression of their performance will hurt their careers. We learned that in program reviews it is possible to commit "abuse of power," but we learned that this is inevitably self-defeating.

We examined the idea of caring for your team personally as a strong leadership skill (or is it a talent?). We also discovered that even if intended as 100% altruistic, such consideration has the benefit of helping the program. The side effect of your caring about the team member has the natural effect of that team member caring about you (and the program). Gratitude is a great motivator. And it has the tertiary effect of "leading by example"—if it is fine to care about people on your program, you can build a subculture of support. Now the team is more cohesive and more dedicated to your program's success.

We talked about dealing with changes, both internal and external. We took some time to think about how the changes affect the dedication of the team and what you as PM can do to use the changes in a positive way or at least to mitigate the damages of "bad" events. We reviewed the natural ups and downs a program and its people endure. Celebrating victories and confronting defeats are both positive leadership activities—and although celebrating is more fun than confronting, fearlessness in discussing uncomfortable issues is important to refine your team's focus and build their understanding of the darker moments in the program's life.

We talked about performance issues and allegations of performance issues and provided some thoughts on how to deal with these difficult problems. (Hiding under your desk was not considered a viable option.)

We learned that relationships with folks you know and how you approach people you do not know are the lubrication of the program engine. People you inadvertently affect the wrong way will not be interested in helping you, even if it is their job. Service folks can easily put your work aside and claim other priorities have taken precedence; they might be telling you that one of the other PMs approached them more politely and respectfully than you did, and they are setting their priorities accordingly!

Finally, we thought about celebrating the program's successful completion. Thanking people is not only the right thing to do; it is also a motivator for them to do more. (Your motivation, by the way, should be the former—the latter is the side effect of doing the right thing.) Putting all your thank-yous in one basket at the "end" of the program is not a good strategy. Intermediate significant milestones are more meaningful and more effective.

Running a successful program is tremendously rewarding. The challenges that you and your team overcome and the good product(s) you produce are great reasons for you to enjoy program management. And the good you do along the way endures and multiplies as the people on the program move on to other opportunities and progress in their careers.

END OF CHAPTER QUESTIONS

For Discussion

1. Do you have any stories about "sticking to the plan" versus "adapting the plan?" Try to find examples of when each approach was valuable or not valuable.

2. John comes to you and says Mary stole his idea. What do you do now?

3. Discuss some of the issues around scheduling, including how to "personalize" the program's schedule and how to build ownership of the schedule.

4. What are some ways that you can build a quality culture within your team? What is the basis for that culture and how do you magnify the natural tendency in most people to do a good job?

5. Your customer has wrongly criticized your team's technical performance. What do you tell them? What would be the approach if the criticism is actually justified?

6. Discuss some of the issues regarding delayed or incorrect information that your customer is required to give you. Perhaps the class has some stories about this all-too-common issue.

7. Describe the "art" of getting stuck engineers unstuck. What do you have to watch out for as you counsel them?

8. Can you use fear of the customer as a factor to encourage progress?

9. Your program is beginning to slip schedule. What considerations should you have as you decide when and what to tell your senior management? How do you involve the team in this "disclosure" issue?

10. Discuss detecting problems by quantitative and qualitative ways. Consider some stories from people who have seen successful and unsuccessful use of each approach, which may be very valuable here.

11. Describe some good and bad approaches to performance coaching. (Have you ever been "coached" and was the technique effective?)

12. Do you have some examples of "scope creep" that has come from the folks on the project? What about examples of customer-inspired scope creep—any examples of that?

13. Can you give an example of when you were offered consideration by one of your leaders and how it made you feel about them (and them as representatives of the company or program)? For example, maybe they encouraged you to take some time off for a sick family member, or came to a funeral for one of your relatives or friends. How about some counter-examples of non-caring—how did these affect your attitude and/or dedication?

14. Is it your job to "protect" the team from negative comments from customers or senior managers? What if the comment is about a particular engineer's capability or dedication? What are the boundaries and the principles behind them for openly sharing information?

15. Your program is coming to an end. What thoughts might be in the minds of the people who are bringing it to conclusion?

Written Assignments

1. Do you have some good and bad leadership experiences? (This is a rhetorical question—we all have these experiences!) Tell us a good leadership story—we tend to remember these less!

2. Is optimism an important leadership trait? Why?
3. Here is a challenge: One of your engineers finds a better approach to an amplifier design and wants to make a change. But the budget for that task is essentially gone. What can or should you do?
4. In one page, not using text from this chapter, describe the major concepts of earned value management (EVM). In another page, describe some of the benefits and pitfalls of EVM.
5. You do not have a requirement for a monthly progress report on your program. Why would you want to write one?
6. What are some of the positive and negative aspects of formal program reviews?
7. Demonstrations of equipment operation can be both good and bad for progress. Describe a few examples of both the good and bad effects of developing demos for customers.
8. Your customer interprets a specification requirement differently from how you and your team interpret it. She had something entirely different in mind when she wrote it. If your interpretation is also valid, how do you handle the situation?
9. Do you use EVM to control costs? "Of course," you say—"Of course not," I say. Why? (Hint: "Control.")
10. You are (fortunately) in the habit of reviewing charging on your program on a weekly basis. You see a new name and do not recognize the budget center. What do you do?
11. In a program review, a functional manager asks you a question that appears to be designed to expose your lack of knowledge on a technical aspect of your program. What might you do in that situation? Give some examples of what you probably should **not** do.
12. Discuss ways to help the team transition between phases—for example, from design to qualification testing—what principles would you employ as the skills needed for the next phase change?
13. The team has just achieved an important goal. It is not clear exactly which members contributed and to what degree. What do you do to get the proper recognition for the right folks?
14. A functional manager tells you that one of the engineers on your program seems to be putting in fewer hours than he is reporting on his time entry. What is your first step? And then what?

8

Claim Identification, Claim Management, and Claim Avoidance

One event that no right-thinking program manager (PM) wishes to confront is the issuance of a claim. This chapter will deal with potential contract defaults, disagreements, and disputes. These issues can result in the issuance of a formal claim against the contracting agency. But more importantly, the chapter will emphasize how to avoid getting into a claim situation. In the ideal world, contracts are clear, specifications are crisp, customers provide government-furnished equipment (GFE) on time, document approvals are per the required response time, comments to your documents are within the bounds of your contractual requirements (not "preferences"), and your engineers are not getting and following informal direction from your customer's engineers. If all that were always the case, there would be far less need for the program management trade. But in the real world, all of those problems can arise, and you must be vigilant to avoid program problems when they do occur. Your program in many cases is governed by a contract that places responsibilities on you, of course, but on your customer as well. Failure to live up to the contract on either side has ramifications. If you fail to perform, harmful things could happen to you. There may be penalty clauses that reduce the contract price if you are late delivering. Or, for various nonperformances or lack of sufficient progress, perhaps your progress payments can be suspended. In a worst-case scenario, your contract could be terminated for cause. All of these potential catastrophes, however, should be in your control, and it is your obligation to avoid them. On the other hand, your customer also can do, or fail to do, things that might hurt your performance under the contract. If your customer fails to live up to the obligations on its side of the contract, it is your duty to first notice and identify the breech and then to take the necessary action to remedy the problem. That action is largely to call the issue to your customer's attention, and ideally resolve the problem—or at a minimum take your own steps to mitigate the impact on the program. But if you cannot make the problem go away, with or without your customer's help, you may have to issue an RFEA (request for equitable adjustment) or even a formal claim. Here are some events/issues that might, if not properly addressed, lead to an RFEA or a claim:

- Your customer is required in the contract to deliver some hardware or material to you (GFE or CFE [customer-furnished equipment]) and fails to meet the date.
- The GFE is furnished on time, but fails to work properly.
- The customer has approval rights on your drawings, and there is a required response to approve the drawings by a certain date. The customer misses that date.
- The customer has approval rights to your drawings, but fails to approve them for reasons that are outside of the contract scope— such as an unspecified preference on how the requirements of a specification are achieved.
- Your customer calls one of your engineers and directs him or her to make some modifications to the design without following processes in the contract.
- Your customer has given you a complex specification, and it is later determined that there was a technical flaw in what was specified, such as conflicting requirements that could not be achieved simultaneously.
- Your customer has given you defective information.

As you can see, this is a long list, but it is by no means all inclusive. And some of these items may have complex technical issues behind them. Furthermore, they can be subject to progressive realization, such that at first, the problem may not be apparent or even visible, but as time goes on, the problem is discovered.

Very often, PMs might start looking for claims when things are going poorly on a program. Surely, there must be some blame on the customer's part for the hemorrhaging of money on your program! While financial problems on your program may alert you to start looking for some shared blame, you want to avoid this situation. It is your job to keep the program running smoothly, and part of that job is to notice customer failings at the earliest detectable time and take action then. If things have gotten so bad because of a customer-inflicted problem, you should have noticed it sooner and taken action then. This is not to say that your failure to notice and take action sooner is a reason not to pursue the RFEA or claim; even if you feel you should have detected and/or taken action sooner, you still have an obligation to bring equity into the contract. But the more vigilant you are, the more likely problems can be recognized early enough to avoid a nasty claim issue.

The following sections are a few examples of this kind of detection and action.

LATE GFE

This is obviously one of the easiest situations to detect. Your contract says that on March 1, your customer is to provide a subsystem for you to integrate into your deliverable for May 1. You are to integrate the customer's apparatus with yours and perform formal tests to verify that they are working together. On March 1, your customer calls and says that she will not be able to deliver the subsystem, would it be OK to deliver it on March 15. You think that you have enough slack in your schedule to still make the May 1 delivery, so can you say, "Sure, no problem!" You can, of course, say OK, but the best approach is not to do so. First, you should not agree to verbal contract modifications. Second, you may *think* that you have sufficient slack in your schedule to accommodate the delay, but that slack was put there on purpose, for your team, in case things did not go exactly as planned. If you were to give up your slack that was due to your customer's delay, you have put yourself in a degraded position. Well, what if the customer tells you that you can have two weeks more on your May 1 date since they are two weeks late getting you the subsystem? Well, that seems fair, so you can say OK to that, right? Not right. Not only do you have the issue about changing the contract verbally as discussed above, but you have created other possible cost and schedule impacts. What about the engineering team that was supposed to begin on the integration on March 1? Would they be 100% gainfully and efficiently employed while they wait for the customer unit? And what about the other end of the task—were those engineers scheduled to be moving on to some other task on May 1, a date that would then be in jeopardy? So, at a minimum, you should not agree to a change in GFE dates without first thinking through all the ramifications and checking with all the affected parties. If it still looks like you can accommodate your customer's request without harm to your project, you will need to get your contracts representative involved and have him or her broker a contract amendment, formalizing the new date and the 2 week slip that you will need to have so that you do not lose *your* slack.

While it may be very tempting to "work with your customer" and avoid the formality of the contract change, in this case it is absolutely imperative that you seek contractual relief. Remember, although you are the PM, and a lot of trust has been put in you and your judgment, you are not the program dictator. If, for example, you made the informal deal with the customer, none of this "private deal" would be in the contract. Any one of your colleagues could identify that the subassembly was late. It would look like you either did not know or did not care, and if your only answer is that you made a deal with the customer, you would be properly "in trouble."

So what could you/should you do in this situation? The best thing you could have done was, as the March 1 date was approaching, to check with your customer to be sure it was on schedule for that delivery. If not, then you would have some extra time to decide what to do and perhaps realign your schedules to mitigate or even eliminate the impact. Your customer will appreciate that you are working together to minimize the problem; a more mercenary or less ethical PM might purposely not mitigate the delay and might blow the delay out of proportion. Working to help your customers, rather than taking advantage of them, especially when they are vulnerable, is one of the best ways to build rapport. But always provide that help "open and above board."

DEFECTIVE GFE

In this example, let us suppose that the customer did indeed deliver the subsystem on time on March 1. Your incoming inspection is performed, and no shipping damage is detected. Your engineers power up the subsystem and everything seems to be fine, but when they integrate it with the equipment your team has designed, the integrated system just does not work. Even worse, it works most of the time, but occasionally, there is a glitch, and your processor stalls. So, is the problem in your equipment or in the GFE? Well, eventually your engineers will figure it out, but in the meantime, while they are investigating, the clock is running, and the May 1 date is getting closer. Furthermore, your engineers are spending time on the problem—time that was not included in your estimate. The worst thing that you can do in this circumstance is to do nothing and hope the problem is resolved "soon." You could check each day, or even each hour, but this is another case where you should not confuse instrumentation of the problem with control of the problem. What you should do is allow a reasonable amount of time to see if the problem is readily solved and then formally notify the customer that you are having trouble integrating the GFE. Tell the customer that you recognize your obligation to make sure your design complies with the requirements, but that the time you spend troubleshooting the problem, if it is indeed faulty GFE, would be considered out of scope. In fact, depending on the contract details, you might need to get formal permission to proceed with the troubleshooting, with an agreement that if it does turn out to be not your fault, that the troubleshooting and resolution time—and any second-order effects, such as other schedule delays—are reimbursable. It is important to take this step immediately upon the time that a customer problem is detected (or even suspected). Figure 8.1 shows a clear example of defective GFE.

Again, many contracts do not allow you to do "unauthorized" work and later expect to be paid for it. While it may seem obvious that the customer would want you to troubleshoot and resolve the problem, you cannot assume that is the case. Even if it is "fair" that the customer should pay

FIGURE 8.1
Defective GFE.

(From Dollar Photo Club, File #4405708 Jjava.)

if a faulty unit has been provided, what is "fair" takes a back seat to what the contract says. Your customer may want to send their own engineers down to help resolve the interface—and that may be a great strategy for you both. It is human nature to assume the other party is at fault, but if you get your engineers and your customer's engineers working together to solve the problem, the solution is likely to be quicker, and the possibility of disagreement as to the cause of the problem is likely to be less.

Let us carry this example to the next step—suppose the customer does authorize troubleshooting and even sends in two of their own engineers to work on the problem with you. (Do not forget to open a new charge number [or budget] specifically to accumulate how much time is spent on finding and resolving the problem.) After three weeks, they find the problem: a communications software delay in the GFE causes an unexpected delay in a handshake between your equipment and the GFE, and the processor in your equipment stalls as a result. Hmm, this could be sticky. The contract is moot on how long your equipment should wait for the handshake, and your engineers have made a reasonable assumption as to how long they should wait, and wrote your software accordingly. Your customer feels that your arbitrary decision of delay time is your choice, that you should have made it longer, and so the problem and its ramifications—time troubleshooting the problem, schedule impact for the delay, and time to "fix" the problem in your software—is all properly your expense.

Do you agree? You should not. The operative principle is that you have every right to expect that all the information you need to perform successfully to the contract is contained in the specifications. As mentioned earlier, the ultimate responsibility for a clear and complete set of requirements belongs to the framer of the specification. Of course, it has to be that way. One must always take responsibility for what one does; and this includes your customer, as the customer is responsible for the completeness of the specifications that were provided. So should the customer pay for all of expenses related to this problem? After all, you were alert enough to put the customer on notice that there was a problem suspected in the GFE and that costs would be accumulated so that in the event it was determined the customer was at fault, those charges would be paid. And, thus, not only is the troubleshooting time reimbursable, so too is the time spent revising and regression-testing your software to accommodate the delay, the time involved in replanning the schedule, and any inefficiencies resulting from tying up your engineers on this issue. So now, the problem is probably properly recognized as pretty expensive. Your customer, however, does not feel that they should pay for everything. After all, you reviewed their specifications and failed to note any deficiency. In fact, they may have included in the boilerplate language (that you accepted) something to the effect that "the Contractor affirms that they have reviewed the requirements and agrees that they are complete and clear." That boilerplate, of course, is included specifically for occasions such as this one. Unfortunately for you, they do have a contractual point. Even if it is clear to the technical people that it is essentially impossible to determine in advance that all necessary information is indeed contained in the specification, you did, in fact, sign up to that unfortunate clause.

It is positive you moved to mitigate impact so that the dollars in question would be as low as possible. And it is positive you have a history of successfully working issues out with your customer, because the only way to resolve this item is by negotiation. Once you both realize that you have some blame in the matter, and admit it to each other, it is just a matter of coming to a fair resolution on cost (and schedule) impact. Do not forget that schedule impact (which is of course a potential secondary cost impact)—level of effort (LOE) program people, such as yourself, may be on the program longer and, therefore, costs for these people will increase. Remember to take steps to keep the team working productively as you work through this issue. Make sure that your contractual coverage is in place for that continued work and thus remains reimbursable, at least in part, when the negotiation is complete.

DELAYED APPROVALS OR CONTRACT ACTIONS

Many contracts have requirements to submit documents and/or drawings for customer approval action. And these contracts also will prohibit (or allow only at the contractor's risk) actions or expenditures that are related to those approval documents until they are approved. These contracts will typically have response times—such as 60 days after submittal—for the customer to take approval action. Your program schedule should allow time while you wait for these approvals. So everything is fine, unless the customer fails to take action within the prescribed period. Absent this approval, you are constrained as to what you may do on the program; at least what you may do without putting your program's and your company's money at risk. For example, you may need approval on an assembly drawing before you may buy components for that assembly. Adding to the complexity, you may want to have all of your approvals in hand before you begin procurement of components so as to get the best quantity-buy benefit. Delays in approval of even one printed circuit board, therefore, can have a significant negative effect on your program's schedule and cost. Even if 99 of 100 approvals have been made on time, that 1 remaining board puts you in a quandary—should you begin procurement on the 99, or wait for that 1 so that you can buy everything at one time?

As you can see, a delayed approval, even on a small part of your program, can have a serious effect. Your job is to minimize this delay, and to seek fair compensation should it occur. Sometimes, a delay in approval occurs simply because, in the maelstrom of submittals, your customer loses track of what he or she is on the hook to approve by when. For these cases, avoidance of the problem is easy—if you are alert. As the date approaches, you can simply call your customer and ask how that approval action on board XYZ is coming. You can offer help by answering questions or providing easy access to the design engineers if there are technical questions delaying the approval. Often, this reminder, even if masquerading as proffered help, will be enough to avoid the delay.

One thing you do not want to do in this situation is to put your customer in a position where he or she is inclined to disapprove the drawing, simply because he or she has not had the time to review it properly. Of course, the customer being too busy is not grounds for disapproval, but since nothing is perfect, your customer may decide his or her least embarrassing solution to the workload issue is to disapprove your drawing and wait for your resubmittal. Even if you turn around the drawing quickly, the reapproval cycle, though typically shorter than the first approval cycle, could again play havoc with your schedule.

If you could "prove" that the customer maliciously disapproved your drawing for an insignificant reason, this in itself could result in a valid claim against your customer. Hopefully, your working relationship with your customer can help you avoid this or even the "veiled threat" of that claim. By far the best approach is to help your customer meet his or her obligations, just as he or she is helping you meet yours. Your customer's success and yours are directly linked and if you both remember this, there can be excellent cooperation in times such as these. For example, if you remind your customer of the impending approval date and he or she says that they will be a week late, you can make it into a claim, or to foster cooperation, you can figure a way that you can work around it with minimal or no negative impact. But because it is a contractual obligation, you should involve your contracts administrator (CA), and either get a formal letter allowing the one week delay or get a purchase order (PO) change notice changing the date formally. Your CA and the customer's buyer can work together on the best approach to mitigate the mutual impact and risk. (Hopefully, they have developed a professional rapport also, and they can work to minimize the paperwork delays that are potentially involved.)

The key point that you should realize from this discussion is that a delay in your customer's contracted action can be a major problem, but you can minimize the impact. An unscrupulous PM might attempt to magnify these problems, and in an adversarial relationship, such things do occur. Hopefully, your program should never get to that state. When your counterpart needs help the most, and you give it to him or her, it is most appreciated. Seeing your customer's vulnerability, and choosing not to take advantage of it, builds trust that can carry you through the next "event."

INAPPROPRIATE DISAPPROVALS OR COMMENTS

Ill-founded disapprovals or inappropriate comments are a first cousin to the delayed approvals issue. Customer approval rights are frequently misunderstood, by both you and by your customer. The customer cannot use approval rights to dictate preferences that go beyond the requirements of the PO or specification. Suppose there is a specification requirement for redundancy in part of your system. You (or your system engineers) architect an approach that meets the requirement. Your customer, on the other hand, had a different topology in mind and would like you to use her vision. Either approach would work, and either approach would meet the specification requirements. You submit your architecture drawing, and your customer calls you and says that she is thinking of disapproving it because she does not like the architecture you chose. Well, what she likes and does not like is of no contractual substance. Her argument may be that her approach also meets the specification requirements; therefore, she is able to direct you to use her approach. This, of course, is not true. If the specification requirement is achievable in any of several ways, then you have the right to design it in any one of those ways that you consider best— best technically and/or best financially. (This will be governed of course by the mission of the equipment and the nature of your contract.) So even if your customer's approach is a little better, or even if it is a lot better, you are under no obligation to meet her preference. For customer relationship reasons, you may indeed want to go with her preferred design, if it is not disadvantageous to you. But she should be encouraged to understand that it is within your prerogative to choose any solution that meets the requirement. If she says that she will withhold approval, or will disapprove the drawing, you may need to remind her of this important rule: approval rights are limited to the bounds of the contract.

You may need to remind her that, as the customer, she is completely able to get what she wants, but it will require a contractual action to narrow the specification to the implementation that she prefers. And, of course, when you receive the new amendment, you will have every right (and obligation) to identify the cost and schedule impact of making the desired change.

Naturally, this is not the outcome that she, you, or your engineers want to see. Again, the solution is communication. The specific approach here could be to involve your customer in the discussions as your engineers are determining the topology. She will have a chance to present her ideas and the reasons behind them. As the customer, her ideas definitely need to be heard and respected, but if your team disagrees, then they also have a right to say so. It would be good for you (or one of your lieutenants who is both customer and contract sensitive) to attend these discussions. Even in the event that a disagreement persists after attempting to work together, at the very least your customer's ideas have been heard, and if a change is to be dictated by a contract amendment, your engineers got to present their arguments for their approach. These meetings, of course, are the ideal place to make sure that everyone understands the rules— the program can choose any of an array of valid solutions that meet the specification. If the customer demands a particular solution, it must be specified. Of course, any delays caused by discussions with the customer that push out the approval of your drawing would be subject to an RFEA or claim. This condition needs to be apparent in the discussions also; this is a contractual way to keep the program moving and to prevent ongoing rethinking of solutions. Using the contract to control the direction of the project is one of the best ways to ensure that progress against the schedule is made.

Refusing a customer request is never pleasant, and here again is where the soft side of your leadership is necessary. You must make sure that everyone understands the rules, including your customer. But you can do it in a nonconfrontational way: "You know Mary, I think several of us like your idea, and we agree it does meet the specification, but we have invested a lot of thought into our approach, and I think you'll agree it meets the specification as well. I wish we could have gotten together on this sooner, but as it is, we need to continue along our current approach because of schedule (or cost) constraints. We realize you are the customer, and we definitely want to satisfy your preferences where we can, but at this point, we would need a PO amendment to change to your approach. Would you like us to give you a formal estimate for that change?"

Depending on the relationships and the experience level of the people involved, you may want to have this discussion with Mary in your office, rather than in front of your team. Mary may be surprised that her only real "power" is to ensure you meet any reasonable realization of the contractual requirements. You (and she) would not like to have that part of the discussion in front of the troops.

The aforementioned is a relatively high-stakes example of this issue, but there are many other less dramatic cases. The customer may issue an "approval with comments" to your drawing. Depending on the contract, this is likely to have the same contractual effect as an approval, but as in the example above, not everyone will realize it. It is also improper for your customer to "force" you to tweak your design with comments, withholding "final approval" until the comments are resolved. You have the right to respond to the comments and include the ones that make technical or financial sense, while politely responding to others as being "beyond the scope of the current contract." Again, it is best to pick up the phone and discuss your position with your customer before putting your position in writing. Avoiding a heated written exchange is also part of your job. By all means, give your customer what he or she wants if it is reasonable and affordable. But you must protect the finances and schedule of your program from inappropriate direction that comes in as comments to your drawings or reports.

NONCONTRACTUAL DIRECTION

Noncontractual direction is also similar to the case described above in that it involves the customer causing contractual impact through inappropriate channels. As such it is RFEA- or claim-worthy. But like all the other instances discussed in this section, the goal is to avoid the situation that could necessitate these negative contractual actions.

One of the primary risks of this scenario (noncontractual direction) is engineer to engineer contact. Certainly, your program's success is enhanced by your engineers talking to the customer's engineers. You would want them to develop a rapport just as you are developing a rapport with your counterpart customer. The best designs will come from collecting and considering all issues, from as many vantage points as possible. Your customer's engineers may know the most about how your equipment will be used in the field and thus represent the voice of the end user. Everyone wants your equipment to provide the best utility to the end user. But how and when these ideas are introduced are key to good program management.

Ideally, all the technical preferences and considerations for the end user are contained in the specifications that are part of your contract. Of course, the "all" is not really achievable; some ideas may seem to be insignificant to the customer as they are writing the specification. Other ideas will emerge as the design is taking shape. Sometimes, your engineer's implementations are contractually correct, but are very undesirable to the end user. For the most part, these issues are properly resolved, without adverse effects to the contract, by your engineers working with the customer's engineers. But in that phrase, "for the most part" lies the risk to your program. It is possible for your customer's engineers to inadvertently direct a change to the contract, and your engineers to inadvertently accept that change. It can be very innocuous:

> *Customer's engineer:* Wouldn't it be better if the A/D [analog to digital] conversion were done in the transmit module?
>
> *Your engineer:* Yes, I guess it might. You're the customer. I'll make the change.

Well, did your engineer ask herself: is there a cost or schedule impact in this change? Probably not—you really do want your engineers focusing first on the quality of the design. But that is "first," not "exclusively." And you do want your engineers to be sensitive to the customer's ideas; "sensitive to," but not "driven by." Some of these changes can be essentially invisible to you and to the other leaders of the program, and all of a sudden, your design may become more complex and more expensive to produce than you envisioned (or quoted). Maybe the "improvement" suggested by the customer is great, but if your engineer incorporates it because your customer's engineer presents it as "direction," you may be in a claimable situation. Again, you never want to be in this place.

One solution that is used is to make sure that all technical meetings are monitored by your CA. This is generally neither an efficient nor even an effective solution. It certainly is costly to your program, as your CA is charging his or her time to your contract as he or she is absorbing about 10% of the technical discussion. His or her ability to accurately detect a scope variance is low—possibly limited to the ongoing question, "Is that idea a scope change?" It is much better that your engineers be alert to this question themselves. As the idea emerges, your engineer and the customer's engineer can consider if there is any scope impact. But to be sure this consideration happens, you must be sure that your engineers "get it" before they engage the customer.

You cannot attend every technical meeting and phone call, but your vigilance can be transported and properly represented by your engineers. They must understand the basic concept of the contract: any solution that is within the bounds of the contract is acceptable, the customer may make suggestions, but they do not have to be followed, and any gray areas need to be investigated before action is taken. It is not generally enough to tell this to the engineers just once. More experienced and seasoned senior folks may genuinely know it, and perhaps learned it the hard way years ago. But your junior engineers are likely to be interacting with the customer (in fact you want this interaction). So be sure that before they do work "unsupervised" that they understand the basic rules of engagement. You may feel like a nag to remind your engineers of these rules before every customer engagement; but it is far better to be a nag than to drift into a claim situation. After you have told them often enough that they say, "I know, I know—I understand about the directed change issue" before you tell them, only then can you stop nagging!

FLAWED TECHNICAL SPECIFICATIONS

In some contracts, the requirements can be very complex. And it is necessarily so, at times. One cannot design high-performance equipment without having that high performance specified! Sometimes the specifications can fill a bookshelf. How can one, in the course of bidding on one of these programs, be expected to completely parse the requirements and verify that they are achievable? In theory, this complete review is frequently but inaccurately assumed to be the case. You are likely to have to affirm that you understand the requirements as part of the proposal process. But actually, everyone knows that complete analysis is not possible for very complex systems and the specifications that govern them. Thus, it is entirely possible that these flaws are detected, not as would be hoped, during the proposal process, but after the award, and perhaps months, or even years into the contract performance.

While you may have had to affirm that you understood the specification at the time of the bid, you may have some recourse for certain items that would be discovered as the work proceeds. Remember that the obligation to provide clear and unambiguous specifications lies in the purview of the framer of the specification. If you are in a situation where the specification requires you to accomplish something that is not achievable, it is time to involve your contracts and/or legal advisors. It is a poor situation for everyone, including your customer, if he or she has ordered something that simply cannot be done. You obviously will have to prove that the particular requirement is unachievable, and that may not be easy—or may not even be purely possible. You may have to settle for the demonstration that you have worked very hard and very intelligently and you consider it therefore unachievable.

There are only two practical solutions at times such as this:

1. Terminate the contract
2. Adjust the requirements

Item 1 is a clear admission of failure—not only for you, but for your customer as well. Remember, that to award your company the contract in the first place, your customer would have to assure his or her management, and maybe a military end-customer, that your company is capable of doing the work. If you cannot, there will be bad press for you <u>and</u> for your customer. The system that you are trying to build may be desperately needed by the end user.

Therefore, Item 2 is the more likely event. Ideally, the adjustment in requirements can be done collaboratively. But the customer will probably need to issue a change notice, and things can get "interesting." The customer may feel that the relaxed requirements represent a decrease in scope, and therefore, if any monetary adjustment is to be made to the PO, it should be downward. You, on the other hand, feel like the flawed specification caused you extra work, work that proved to be useless toward completing your obligations under the PO. You feel like you should be paid for those efforts, and the PO value should be *increased*, not *decreased!* Some serious negotiations could therefore be expected to follow.

In these negotiations, you will be most effective if you understand your customer's position and even sympathize with the customer, certainly on a personal level, but also on an "official negotiation" level. Remember to think "win–win"—how can you both come out of this problem with the least negative impact? And how can you preserve the collaborative spirit that you have worked so hard to build?

This is certainly a challenge—funds that were wasted on the unachievable are just plain gone, and there is perhaps not enough money in the contract to complete the work. Even if you were to forego your profit, which, by the way, you do not want to do, there still might be insufficient funds to complete the work.

So is there a remedy? When two intelligent parties work together to solve a complex problem, the answer is usually "yes." It may require some out-of-the-box thinking, but talking through possibilities with your customer is a great start. One possibility is to find features that your customer had wanted to have but does not actually need to have to constitute a successful system. If the customer were to eliminate these requirements, then the money saved could be applied to this recovery effort. It would probably be better to have the customer ask you to price the change as one package: "Please provide a proposal to relax specification A and modify feature Q and R as follows."

Depending on the contract and the content of A, Q, and R, this approach may be either clever or improper. Your contract personnel should be an intimate part of this recovery. Although they may not understand the technical issues and, thus, may not be very able to contribute ideas, they will know what can be done with the PO, within both the letter and the spirit of the contract.

Failure to reach an agreement may logically result in an RFEA or a formal claim. As discussed before, this is a very undesirable "solution." What really happens is that it is now necessary to negotiate anyway, but much of the freedom that could be brought to bear is lost in the formality of the claim process. Your customer is at least as anxious as are you to resolve the problem "peacefully."

DEFECTIVE INFORMATION

This scenario is similar to the situation discussed earlier, but is much easier to handle. The situation is much more straightforward: information provided by the customer to you is wrong. On the basis of this incorrect information, you have expended effort which has proven useless.

While there is no need to share the blame, the mutual motivation to work to a mutually satisfactory solution is still there. In this case, you may be "holding all the cards," but you must still work to a win–win outcome. In fact, this negotiation may be, ironically, a little more difficult. Since the customer is responsible for providing the incorrect information, the problem is clearly his or her's. If your customer feels that there is potential embarrassment or even a career-harming action because of the error, negotiations may take an oddly confrontational tone. It is good to recognize this as the motivation and work all the harder to minimize the problem, and to assure your customer that, although you know you are holding all the cards, you intend to play them to your joint betterment as much as possible. Engage in conversations along the lines of—"Well, this may be a tough problem to get through, but let's sit down and think about what we can do." If you really do care about the success of your customer as well as your own, the collaborative nature of your approach should be evident.

At this point, it should be clear that your prime mission in the face of customer-induced problems is to work to minimize the size of the problem and to work with the customer in doing so. You need to protect the integrity of the funding and the schedule on your contract, but you also need to help the customer out of what could be a very bad situation for him or her. And a very bad situation for your customer will undoubtedly be a very bad situation for you and your program. You do not have to be sympathetic to your customer because you are "nice"—being sympathetic is actually quite self-serving. Remember, if you help someone when they are in trouble, they will remember it and will hopefully do the same for you.

But of course, not all problems can be handled satisfactorily and properly within your and your direct customer's purview. And although you may understand your contract very well, and be somewhat familiar with the law and the terms and conditions of the order, you are not the expert, nor are you expected to be. It is thus very necessary when these situations arise to bring them to the attention of your CA. Because of the high stakes involved, you may want to be sure that the situation is documented and presented to him or her in writing, or at least in an email. In addition to protecting you from any future allegations that you did not react fast enough or clearly enough to situation, putting discovered issues in writing gives better assurance of understanding on your CA's part and allows him or her to more easily seek guidance from his or her manager or from the legal department.

It is likely that your CA will also be in favor of a win–win solution, but do not be surprised if he or she is not as driven to this goal as you are. Your job is to get the job done and protect the company, pretty much in that order. Your CA probably does not care that much about the successful progress of your program, but does care, and is paid to care, about protecting the company. You may have to help him or her understand that the situation is best handled as a "win–win"—no matter how strong your position is. The same applies to your senior managers. There is a risk that you might be considered a "customer-lover"—and actually you should be, and should wear this intended slur with pride. Or they may think you may lack the stamina to stand up to the customer. People with adversarial tendencies tend to expect adversarial behavior in others— and when they do not see it, they may infer that the lack of confrontation is due to cowardice. As people get to know you to be driven to win–win solutions, this incorrect inference will likely disappear, but you should be alert to it. Just as adversarial people are slow to recognize "win–win" players, so too are win–win players slow to recognize adversarially minded colleagues.

Be that as it may, and possibly because of this tendency, you may be reluctant to bring in outside "help." It is great that you feel confident that you can bring the issue to a satisfactory conclusion without involving your contracts or legal colleagues. But it is dangerous, if not outright improper, for you to attempt to manage these problems without consulting them, and consulting them at the earliest possible moment.

If you have a particularly good rapport with your customer, and if he or she has helped you out of uncomfortable positions in the past, it might be tempting to try to sweep some customer nonperformance issues "under the rug." But there is a great danger in attempting to do so. If you notice a nonperformance and fail to take action on it, you may compromise your program's (and your company's) position and make recovery of attendant costs more difficult or even impossible.

Thus, this is a very difficult situation for you to manage. On the one hand, if you are too quick to take action, you can make the problem worse a mountain out of a molehill and you can alienate your customer. On the other hand, if you delay and "hope for the best," things can get much worse and it would be your fault for not taking action sooner.

The key, as with most of your problems and potential problems, is communication. If you are not sure whether something should be brought to your CA's attention or not, err on the side of conservatism, and tell him or her.

CLAIMS AGAINST YOU

On some programs, especially on large ones, it is possible that your program may want to subcontract some aspects of the project. Subcontract management is discussed further in Chapter 7, but it is addressed here due to the potential for your subcontractor to issue RFEAs or claims against your company (through your program). The same issues discussed earlier that can be the substance of your claim against your customer, can also be material for a claim of a subcontractor against you. The best protection is to be sure that the claim's causes are avoided by you and the colleagues that deal with your subcontractor. You may have a subcontracts manager on your team, and if so, he or she would be expert in avoiding these issues. But on smaller programs, or even large programs with small subcontracts, you may actually serve as the de facto subcontracts manager. Your buyer may hold the paperwork and the actual authority to place POs and amendments, but it is likely that you will understand the subcontract issues (and what you expect to get from your subcontractor) much better than does the buyer.

Simply, to avoid trouble, invert the claims scenarios discussed earlier. Manage your subcontracts so as to do the following:

- Provide any material you need to provide to your subcontractor ON TIME.
- Make sure the material you provide is correct and functional.
- Respond promptly to submittals and questions. Be sure the contract has clear "turnaround times" and that you and your team achieve each and every one of them.
- If approvals or quality inspections are part of the order, make sure the time requirements are clear and that you meet them.
- Brief (not only once) your team to ensure they understand that they are not permitted to give contractual direction (or anything that can later be construed as contractual direction).

- Make sure the requirements of the subcontract, especially the technical requirements, are crystal clear. Suggest a meeting with the subcontractor, if warranted, to assure clarity.
- Finally, make sure that the information you give the subcontractor is correct and complete.

As the framer of a subcontract, you must think like a customer. Although you may be working to establish rapport with your subcontractor, it is likely that that relationship will be more tactical or episodic than your relationship with your own customer. This means that your subcontractor may put profit ahead of relationship and may be more likely to exploit you errors or omissions. While most of your subcontractors are likely to act ethically and reasonably, it is important that you plan for their not doing so. Make sure your actions in support of that subcontract are of high quality, are done on time, and are well documented.

OTHER CONSIDERATIONS IN CLAIM MANAGEMENT

By this time, it should be readily apparent that good communication with your customer is the best protection against needing to execute an RFEA or a claim. Warning the customer of his or her upcoming obligations (such as approvals); seeking clarifications for unclear specifications—and (politely) asserting your right to any valid interpretation of those requirements; identifying early any possible performance problems with equipment that he or she has provided to you, etc., are best practices. You can call frequently or send him or her emails regarding these issues, or another approach to consider is the creation and transmittal of a monthly progress report. In addition to the convenience to you and your customer of collecting progress news in one place, and in an easy-to-read format that is regular each month, it has another purpose as well. It provides documentation of when you notified the customer of issues and can contain and help manage a list of action items for both you and him or her.

Although this report may not have any contractual force, it is, in addition to being helpful as discussed, a great low-key record of progress and issues. This could be very important if relations were to go terribly wrong on your program (for example, your friendly customer engineer is replaced by someone neither knowledgeable nor helpful). Having things "sort of" on the record will protect you from selective memory and will keep visible any and all customer actions you need to proceed in the most effective way.

Improper Use of Claims

It is important for you to know that claims can sometimes be used inappropriately and unfairly. While you and your company may be too ethical and honest to act in this manner, you should be aware of them so as to protect your company from unfair completion. Suppose your (potential) customer issues an RFQ (request for quotation) with an obvious flaw. The correct and ethical thing to do is to bring it to your customer's attention and allow him or her to modify the RFQ, correcting the error. But suppose someone else wanted to win the bid in any way possible. One strategy that an unethical competitor may employ is to observe the flaw in the RFQ and say nothing about it. Knowing that there is a clear fault in the RFQ or specification, he or she may decide to enter a very low bid and capture the business. Once captured, he or she could point out the error at any point. This would force the customer to issue a change, and your competitor would then have an opportunity to bid on the change. While legitimately, only charges related to the correction should be included in the bid, very often it is not possible for your customer to accurately distinguish the legitimate additional costs from those that are "extra"—and intended to make up for the intentionally low bid. The result is that you lost the bid on an unfair basis, and your customer may have a contractor who is taking improper advantage of them. This situation could be the basis for a formal protest. While it is extremely difficult to prove that your competitor had an "evil" intent, your recognition of the possibility should be reported up the line, including to your legal department, where experts in these matters can decide on a course of action.

The best protection from this kind of inequity is through a thorough review of the customer's RFQ and bid specification. Your competitor cannot exploit an error in the bid package if you detect it and bring it to your customer's attention during the bid process. You may be surprised to learn that all companies and all proposal managers are not as ethical as your company and you are. Be assured that not all are ethical. You must remain vigilant to any improper actions on their part, and, in this case, preclude them by careful review of the RFQ and technical specifications.

SUMMARY

Your contract is not perfect, and your customer's performance of his or her obligations may not be perfect either. These imperfections can cost your program time and money—and it is your obligation to prevent them from occurring. The best way to effect this prevention is through a thorough review of requirements during the RFQ. And then after award, you can prevent claim activity through frequent and collaborative discussions with your customer. Your communications should all have the theme of helping each other and "keeping each other out of trouble." Good communications and honest intent can prevent a multitude of problems and can keep a problematic specification from costing undue amounts of time, money, ill will, and unnecessary controversy.

END OF CHAPTER QUESTIONS

For Discussion

1. Claims are a great way to increase the dollar content of your program—true?
2. A scope issue arises on your program, but you want to keep making progress. Your customer says, "Go ahead, we'll work it out later." How dangerous is this situation, and what should you do?
3. One of the most challenging issues in customer relationships is where to "draw the line" for gray-scope areas. What thinking should go into your line-drawing thinking?
4. One of your lead engineers has a very good rapport with a customer's technical lead. You are worried that his desire to please his "friend" might cost your program in schedule and money. What should you do?

Written Assignments

1. Discuss some strategies on how to reduce the probability of needing a claim on a program.
2. If you feel that a scope issue may arise, what are some of the immediate steps you should take in case the issue gets bigger rather than smaller?

3. Your customer disapproves one of your drawings because she does not like the way you implemented the requirement. What discussion do you need to have with her? Do you think you might need to include your CA—and if so, when?
4. You have been working on a program for a year when you find that one of the requirements is physically impossible to achieve. Is this a potential claim situation? What if you signed an agreement that said you understood the contract requirements?
5. You subcontract part of your program to a small firm. Things go badly for the company, and you hear a rumor that they are preparing to issue a claim against you. What do you do?
6. Can a progress report help you avoid claims? You say yes, and you are right; but, remember it is not part of your contract, so, how can it help?

9

Leadership Models

LEADERSHIP: GETTING PEOPLE TO DO WHAT YOU WANT THEM TO DO

That seems like a good working definition, but it leaves out the *how*. And it is simplistic on the *"what."* An effective program manager (PM) may have a great understanding of the strategic goals of his or her program, of course. But the tactical ones, the way that things must happen in order to get to the strategic goal, are often beyond his or her expertise. So, effective program leadership is not getting people to do what *you* want them to do; it is getting them to do the right thing for the program. Only a PM who is both arrogant and uninformed would think that he or she alone knows all that is necessary to make the program successful. An effective PM knows that he or she must enlist the talent, intelligence, experience, and initiative of the team, in order to do not "what you want them to do" but rather "what should be done."

The How of Leadership

And now, for the "how" of leadership.

This chapter is entitled "Leadership Models" rather than "Management Models." The concept of "manager" involves power and authority. Effective PMs recognize their "power" is much less important than their leadership. In most organizations, PMs do not have tremendous power anyway. For example, they cannot hire and fire people at will, or promote them or give them raises. Therefore, while PMs are often thought to be "powerful" that is more perception than reality. The real "power" of the office is rooted in leadership skills.

The functional managers (FMs) in fact have a great deal more power than the PM in terms of being able to positively or negatively affect an engineer's career. Since for many engineers, career and salary advancement are important motivators, it is good to form alliances with the FMs to help motivate engineers from that standpoint. It is good to have "powerful" friends. But in my experience, you should choose influence over power, even if you had the choice! Remember, part of your PM job is to attract the best and smartest people to your program. Figure 9.1 shows two leadership environments. In which environment do you think talented people would choose to work? And remember, smart people are good at making good choices.

The PMs control the program's purse strings so maybe that is where the power lies. Maybe—but it is pretty limited; folks know that even if it is an unspoken realization. The "power of the purse" is a good concept in an open air food market, but does the model apply in an organization where work has to get done? Suppose the PM is dissatisfied with the work that the software team is doing. It is not going fast enough, and he or she is thinking of throwing his or her weight around. (Sounds a little like the PM is more angry than anything.) So, he or she cuts off charging and makes those folks idle. Sure, they have stopped charging the program's budget, and sure, their FM has a big problem in finding them gainful work to keep them off overhead charges. That does sound like power, does it not? However, the PM would be taking that action to get the work done faster. *Stopping* the work because it is not progressing quickly enough is a very poor solution to the problem. Everyone knows that, so threats to cut off charging are immediately seen as empty. Threats are poor motivators in the first place, but empty threats are laughable and 100% counterproductive. It is not as though there are 10 software engineers sitting in the lobby,

FIGURE 9.1
Work environments.

([Left] From Dollar Photo Club, File #488503089 Orlando Florin Rosu; [Right] from Dollar Photo Club, File #49968512 snap galleria.)

ready to pick up where the current team has been working. The PM should never get into that situation, where he or she has only empty threats to attempt to improve performance!

Rely on Influence, Not Power

So with limited real power, the currency of the PM is not money; it is influence. The effective PM uses that influence to motivate the engineers. Primarily and fortunately, the people on the program are often high on the self-motivation scale. If you are an engineer by trade, you remember that in college the engineers were the folks doing problem sets and staying late in the lab while students of other disciplines seemed to have a great deal more time. The discipline that these serious students showed then is still part of their competencies, skills, and work ethic. It really is not that difficult to get people who like to work, and expect to work, to work! And furthermore, you probably work in a company that can be very selective toward those they hire. Aerospace and defense companies attract talented people who want to do important and challenging work. Because there are more people who want to work for your company than you can possibly hire, you can select the best of an already hard-working, self-motivated population. So maybe you do not need power to get the work done. Thus, your influence, not your power, is the tool you need to make the program successful. This is great news, because how much power you have is out of your control—but how much personal influence you have is absolutely under your control.

So, where does that influence come from? Think about people that influence you—especially those that influence you to work hard and do your best. You are likely to realize that you are influenced by those whom you respect—and so are the people on your team. Gaining the respect and trust of your team, therefore, is the best use of your time and effort, and is the mark of a true leader. You do not have to be a behavioral scientist to understand the motivation of your engineers. You only have to look inside yourself.

We mentioned in Chapter 7 that leadership is an ephemeral quality. And because many PMs are more comfortable with the tangible, leadership is not really closely examined. Many people may pay lip service to the need for leadership, but few people take the time to think about it. There are many theorists of leadership (for example, Maslow, Herzberg, and McGregor), but as engineers, we are by nature more interested in application than theory. Let us use that engineering process of applying theory to practical use and look at leadership that way.

Examining Leadership Theory

McGregor's Theory X and Theory Y assume there are two basic types of workers:

> X: Workers are lazy and constantly need to be told what to do.
> Y: Workers are self-motivated and generally know what to do without being closely directed.

While neither extreme is completely correct all of the time, Theory Y is generally a much better model of engineers, and in fact for all the folks working at your company. (Distinctions between managerial, professional, salaried, and hourly workers are unimportant as long as the people are treated fairly with an expectation of good behavior.) So, if you have large population of Theory Y folks, it is not necessary to monitor them and micromanage them. Thus, your role becomes more centered on creating an environment for collaboration and coordination. You need to help the team achieve the strategic goals of the program while you trust them to work though the tactical details that you may not even understand.

One of the misunderstandings of leadership is that the term "leadership" is used differently for different circumstances. The military teaches leadership, but that style of autocratic non-questioning of orders is not really what you want in your team. In fact, you want to minimize "orders" and maximize "agreements." It is a different quality that causes soldiers to run into harm's way than the quality that causes engineers to want to sit down and collaborate. Infantry platoons might not be successful with participative leadership:

> *Sergeant:* Do you think we should run up that hill?
> *Soldier 1:* There are enemy soldiers with guns on that hill! That doesn't seem like a good idea, Sarge.
> *Soldier 2:* Look, there's another hill, without enemy soldiers. I suggest we run up that one!
> *Soldier 3:* Run? Heck, we can *walk* up that hill!

To win wars, soldiers often need to do what they are told, quickly and without question. On your program, however, you want your team to not only be open to alternatives, but, in fact, to create them. The kind of leadership that involves multiple points of view and empowers contributors of new ideas is not the same kind of leadership that motivates soldiers to run up the right hill. You really do not want blind obedience to your orders; you want the minds and hearts of your team engaged in working toward the best decisions. This comes from empowering the team and practicing participative leadership. In complex situations where everyone has different information and backgrounds and everyone is working to get the best design, participative leadership cannot be beaten:

> *PM:* Do you think we should put the processor on the front-end board?
> *Engineer 1:* Sounds good. It would save space.
> *Engineer 2:* Yes, it would, but I would worry that there would be noise on that front-end signal. You know, we are only dealing with microamps.
> *PM:* Good point. Let's have a separate board. Everyone agree?

So in a situation where you want to draw upon the knowledge and experience of everyone on the team (or in the meeting), the PM's role is to lead the discussion and drive it to (ideally) a consensus decision. The effective leader remains patient as people argue their points and ensures that the discussion does not turn negative. He or she makes sure the less vocal participants are heard. Very often, the people with the best ideas are "quiet types"—they may be quiet because they like to think through things thoroughly and are reluctant to present ideas that are not fully investigated. You may have to wait for them to present their ideas. As the leader of the discussion, you may also want to contribute ideas and make suggestions—but do so as a member of the team, not as the "boss." In these meetings (or even in email transactions), the PM's role is to make sure the group gets to the best decision with true consensus, but as quickly as practical.

This type of leadership, where the PM is the catalyst of the decision, will produce the best solution. Of course, the PM must be humble enough to appreciate that the collective group, and probably several individuals in the collective group, are smarter than he or she is!

Thus, if you are going to be an effective participative leader, you must have humility, and you must have the respect of the team. People will follow people that they fear or people they respect. We have already ruled out the fear motivation for our daily work. Intimidation works for the instant, or if the fear level is sufficiently high (fear of death or physical injury) it can work for a while. But autocratic leaders are limited by their own knowledge. People will not tell the autocratic leader when he or she is wrong, either because they fear him or her or because they do not care about his or her success. Dictators' inner circles are filled with yes-men and yes-women. Ideas not in line with the dictator's can get you shot (or fired).

New ideas and insights are thus suppressed, and catastrophe caused by this suppression of information is guaranteed to follow.

> "The peasants have no bread!"
> "Well, then let them eat cake."
> (*Suppressed information:* "If they are starving, you are in big trouble, Your Majesty!")

A little participative problem solving could have spared a few leaders' heads!

Not all program situations, of course, lend themselves to getting the whole team together and making a decision. But important matters make the collaborative effort very worthwhile. It is highly advisable, for important decisions, to gather all the affected people and work to the best solution. But such consensus is not possible, or even necessary, for the dozens of decisions that the PM has to make during the course of a day. Since full-time participative leadership is not possible, it begs the question, "What do I have to do to be an effective leader, in general?"

Achieving and Maintaining Respect

We mentioned earlier that the primary need of a nonauthoritarian leader is respect. This is obvious in our daily experience. When a political leader is found to have done something dishonorable (too many examples to choose from), he or she loses the ability to lead. Even the Commander in Chief, who has plenty of power, loses much of his ability to lead if he is found to have done something dishonorable. How much more so for the PM, who does not command the Army, Navy, or Air Force?

Think, then, of the leaders you know and respect. Think of the traits that they have in common. We do not usually think of "why" we respect or do not respect our appointed leaders, but we maintain a little list of standards in our heads. Subconscious processing of this list against what we observe allows us to consider someone a good leader or not. Chances are your internal list and that (internal) list that each member on your team has are pretty similar. On most lists you would find the following:

- Intelligent
- Competent
- Courageous
- Fair

- Insightful
- Sensitive (although many may not admit to having this on their list)
- Caring (same here—too touchy-feely to talk about)
- Experienced
- Attentive (listens to me and my ideas)
- Inclusive (makes sure my quiet buddy gets to give his or her opinion)
- Humble
- Confident (when things get tough, we do not want our PM to panic)
- Optimistic (we do not need leaders who predict failure)
- Hardworking
- Trustworthy (follows up on promises and agreements)

Intelligent team members respect these traits in their leaders because they know these traits lead to success. Good leaders make sure that these traits are always on display. Not all the members of your team see you all of the time; therefore, you must constantly exhibit these values to be respected by your team.

Since it is a process, encompassing several days, weeks, or months, to get the respect you need to be an effective leader, it can be risky that your program will be well under way before you acquire the respect you need.

Fortunately, most people will give you the benefit of the doubt until they get to know you. As they do, and as you exhibit the respect traits that are on the unwritten lists that they keep in their heads, the better your leadership abilities will become.

As luck would have it, however, it is much easier to lose respect than to build it. Just ask the indiscrete politician who gets caught doing something improper and/or humiliating. So as you are building your "leadership image" with the team, here are some things that can set you back to ground zero, or worse:

- Losing your temper and demeaning a fellow employee.
- Acting dishonestly—on the contract, or even when it is time to share a lunch bill.
- Leaving the team in the lurch—a report has to go out and 10 people are staying late to make sure it goes out that evening. You leave on time because you are tired.
- Arrogance—or elevating your position.
- Showing favoritism or nepotism.
- Failure to listen to someone who has something important to say.

- Self-serving ambition.
- Chronic unavailability to your team.
- Chronic indecision.
- Chronic inflexibility.
- Prejudice toward people not like oneself.

Certainly some of these failures are more severe than others, and like the "respect traits" everyone is entitled to his or her own private list. People who are themselves overly ambitious would not be as disenchanted with your self-career-focus as people who are not. Showing favoritism does not seem so bad if you are one of the favorites. And fortunately, no one expects their leaders to be perfect. So when (not if) you show a failing, the most important way to recover is to apologize to the people affected. (And you will need people who respect and trust to you to help you realize when you need to apologize.) Leaders who lose the respect and confidence of the team will be replaced—either by administrative action or by de facto disenfranchisement. A PM who no one listens to is a drain on the program—no longer its leader.

In Chapter 10, we will focus on communication, but for our consideration of leadership traits, we recognize that the leader can transmit information and receive information. Certainly, the PM must do both, but to be a good leader, you need information, and you can get information only when you are not talking. It is fortunate that the same behaviors that make you a good leader also make you a good listener. And the good listening that you do gives you information to be a more effective PM.

Finally, remember that leadership requires courage and confidence, and with these comes the ability to take personal risks. A leader cannot get a group to move in the same direction by "blending in" and "not making waves"—and it takes courage to be different. The herding instinct is more prevalent in human behavior than people realize. In your next meeting, observe how your colleagues exhibit the sheep syndrome—keeping quiet, even if something should be said—relying on others to say it. Observing how fish behave can provide another good model of leadership. Fish form schools for protection—not protection of the school, but protection of the individual. The safest place in a school of fish is in the middle—fish on the outside are more likely to be eaten. Thus, the survival instinct drives all the fish to seek the middle. That is what forms the school. But, at any one time there is one fish who is not as anxious to survive and get to the

FIGURE 9.2
Leadership in nature.

(From Dollar Photo Club, File #4731674 Bernd Ege.)

middle—that fish is in the front and leads the school. Direction of the school is chosen by that "leader" who has to be a little less interested in self preservation and a little more interested in getting the school to go in the right direction. It is inherently riskier to be that fish! Figure 9.2 shows the courage and vulnerability of that lead fish!

SUMMARY

Effective PMs are influential and achieve that influence by earning the respect of their team. A team formed by a new PM will likely assume that he or she is a good leader but will silently be assessing if that PM is worthy of their trust and respect. Team members will respect leaders who consistently demonstrate the traits that they feel leaders should have. Since developing that respect is a process, not an event, good leaders are constantly attentive to showing the positive traits that engender that respect.

The accumulated respect, however, can be lost in a moment of indiscretion. Leaders may not have to be perfect all the time, but the loss of respect-capital that is due to one outburst or one rude remark can be nearly fatal. Always act in the way that enhances the respect of your team—and if you fail, apologize sincerely, being sure all who saw your temporary failing hear it.

END OF CHAPTER QUESTIONS

For Discussion

1. Do you think the people on your program are out to make the program successful or out to make themselves successful? Or a combination of both? Or maybe there are some other motivations?
2. Discuss participative leadership in various circumstances. When might it work and not work?
3. Are good leaders "humble"? How this is possible if one needs his or her leadership to be "confident"?
4. Name some famous leaders in history. Were they participative, ruthless, or charismatic? Is "fame" always consistent with success? Discuss whether some "famous" leaders were actually figureheads of successful enterprises and not the cause of the enterprise's success (or failure).
5. Someone forgot to order a critical part for a breadboard. It costs $100. You think someone should use his or her personal credit card to buy the part quickly—but they might not get reimbursed. Your action is…?
6. If you cannot motivate people by reason, it is OK to motivate them by fear—or is it? Consider the application of this thinking in terms of success, apparent success, failure, or apparent failure?

Written Assignments

1. Despite the fact that you ascribe to the theory of always treating your colleagues with respect, one day you really "lose it" and shout at a member of your team. What should you do? What if others witnessed your lack of control?
2. Discuss what is good and what is bad in this definition of leadership: *Leadership: Getting people to do what you want them to do.*

10

Communications

"Why wasn't I informed about this?"—the plaintive cry of a manager (not a leader) who has just found out some disturbing news. The true answer is, "Because I didn't think it would help!" That is not usually the "right" answer, because the question is more accusatory than information-seeking. Often, the manager asking the question is angry—and we all have learned that it is a dangerous thing to engage an angry person, especially if he or she is your boss!

When things go wrong on a program, very often it is blamed on poor communication. And very often it is true. Communication failures are one of the basic causes, or at least why things get worse instead of better. The stock solution to this age-old problem is often an increase in the type of communication that is shown in Figure 10.1.

Chances are that the problem did not occur because the people on the program were not familiar with the strategic plan of the organization. Chances are that the people who saw the problem forming attempted to "put the fire out themselves" instead of calling the fire department (the program manager [PM] in this case).

A schedule that is about to implode does not benefit from any kind of lecture from anyone. Schedule performance improves by determining what is wrong and then taking action. Your lieutenants on your program know that, and every day, without involving you, they solve problems. It takes a particular talent on their part to know when to alert the PM that trouble might be brewing. What you want them to do is let you know when things might be going wrong. Why would they or why would they not tell you? That is the crux of the communication problem.

Figure 10.2 illustrates the better remedy to a communication-failure diagnosis.

FIGURE 10.1
Communication, in a way.

(From Dollar Photo Club, File #6018182 popaye.)

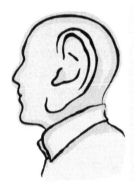

FIGURE 10.2
Communication, a better way.

(From Dollar Photo Club, File #25798626 robodred.)

If your lieutenants saw you as "all ears," they would naturally tell you about what was going on, including brewing problems. Notice that the PM in the illustration is remaining calm. He hears a lot, but is still calm! And when he hears disturbing news, he does not lash out at the person who brought him the news. Killing the messenger has pretty much gone away with the ancient Greeks, but negative reactions, including anger, in front of the messenger—well, we all may not have evolved to the point where that is no longer a concern!

Instead of getting angry, the effective PM says, "What can I do to help?"—and means it. His or her reaction is not judgmental and but instead is constructive. This makes it easy for the team to talk things over with him or her. This is the very important trait of approachability. Your team must <u>want</u> to talk to you, not try to avoid you!

People will tell you things only if

1. They think it is in the best interest of the program.
2. They think it is in their own best interest.
3. They think you care.
4. They think you will help them do something about it.

and they will not tell you things if

5. They think it will "make trouble" for them or their friends.

Reason 5 can easily trump reasons 1, 2, 3, and 4 put together.

Suppose for a moment that you are one of those authoritarian managers. You hear of a schedule slip (maybe you were told by an unsuspecting team member, or maybe you eavesdropped at the watercooler), and you decide to take action! Your actions might include:

1. Mandatory overtime
2. 6:00 a.m. meetings with the team
3. Getting everyone together and conducting the Grand Inquisition
4. Going to the functional managers (FM) of the people involved and complaining
5. Changing the staff and/or the assignments
6. Demanding a recovery plan that should include some of (1) through (5) above.

If the folks on your program think one or more of these would be your reaction, *why would they* tell you any bad news?

This fear of something bad happening when the fire alarm is pulled is exactly why it is pulled too late or even not at all. Sometimes it has to get so bad that the problem is visible even outside the program. If the problem is that big, recovery is not going to be easy or pretty. You need to know when someone smells smoke, not when the flames are leaping out the second story window!

Actually, you want to be told when your lieutenants *think* they smell smoke. If you know there might be a problem brewing, you and the involved parties can work together to decide what to do about it.

It can come to pass that the highly nonapproachable manager is starving for information. If this is a chronic problem, he or she will develop wrong-headed communication strategies to get information. Discussions with this type of manager can resemble interrogations rather than conversations. He or she feels that you are withholding information (or maybe even lying!). Why would they think that? Easy—because in order to prevent trouble, other less forthright staffers have been avoiding telling that manager bad news, and because they fear the interrogation, they avoid telling him or her <u>any</u> news at all. In fact, folks will avoid walking near that manager's office to preclude even casual conversations. The result is information starvation—which the authoritarian manager attempts to resolve by more aggressive interrogations! This is a communications downward spiral that does no one, including your company, any good.

Good, fear-free, open communication is the hallmark of an effective team—and that communication is within the team as well as with its leaders. If the environment is open, good communication can be the norm at all levels. The effective PM makes sure that people are not afraid to talk to him or her!

COMMUNICATIONS AMONG THE TEAM

Of course, no matter how approachable you are, there will be information that you do not need to know and that you could not possibly understand or use. Information has to flow among your team in an efficient way, and it has to flow outside the team for centralized services such as shipping and drafting. Figure 10.3 shows your team talking among themselves.

You are in the middle of the picture, slightly larger than your team (I guess that ego is not completely in check!). But you will notice how few of the communications paths come to or through you. So then is your communication obligation limited to just your personal I/O (input/output)? Not at all. Good communication within your team is your responsibility as well as the responsibility of the people on the team. If information does not flow freely among the people who need it, the team's progress will be stunted.

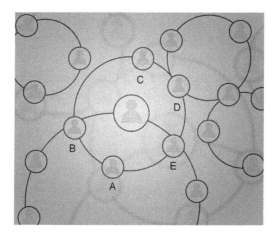

FIGURE 10.3
Team communication.

(From Dollar Photo Club, File #53501896 backgroundstore.)

Diagnosing poor communication among certain members of your team is difficult. Even if your approachability coefficient is high, you are not likely to hear of personal grievances or lack of respect between members of your team. This is why it is so important to build your team out of people who have mutual respect for each other, or at least are willing to assume the best about a teammate they do not yet know. However, it is not always possible to ensure that everyone on your team is "just right" in every way. FMs may have the ultimate responsibility to make assignments, and it may be difficult for you to get exactly the folks you and the rest of the team want. They may not be available, or they are promised to another program. Thus, you may have some people on your team who do not communicate well with each other.

Notice in Figure 10.3 how convoluted the information flow would become if person A did not talk to person B. It would take three or four people to get the information where it needed to go. It is an extreme case that two people on your team just will not talk together, but that is one of the reasons that it is so hard to diagnose this type of communication problem. A and B may appear to work together, but either or both of them may work to avoid contact—and information flow.

How, then, do you diagnose this kind of problem? Hark back to Figure 10.2. If you keep your ear available, you will definitely learn more things about the interactions of your team. But you do not have to be passive about it. The PM's favorite question—"How's it going?"—can yield a wealth of information if you take the time to listen *and* to listen for innuendos. If you suspect a communication problem, you can up the ante a little by saying to B, "How is it working with A; is she giving you all the help you need?" If B is really resentful of or belligerent to A, you will probably hear it in B's answer. It may be subtle, because people do not like to "make trouble." But the tone of the answer, and maybe some gentle follow-up, can at least identify a concern.

So once diagnosed, what do you do about it? You should talk to them individually and remind them how important it is (and in fact part of their job) to get information to where it needs to go. You should make them part of the solution—they may have suggestions on how to resolve the issues that are compromising their performance, or they might recommend an adjustment in assignments. As in all problem solving, the folks involved should own the solution. The key is not to allow the problem to persist but to find it and take action. What action you take is part of the art of program management and another test of your leadership ability.

WHAT ABOUT COMMUNICATION OUTSIDE THE TEAM?

You and your team members have to work with the rest of the organization to get things done. You will need IT services to get the new computer loaded with the right modeling software. You will need Incoming Inspection to expedite receipt of lab material in order to get the prototype working. Security will have to certify your software room. When you communicate outside the team, you often will not have the rapport and friendliness that (hopefully) exists on your team. Also, while you and your teams are extremely committed to your program's success, other parts of your organization are servicing others' needs as well, and those PMs think that their job is the most important one in the world, also! As usual, respect and humility go a long way, and emails that clarify your specific requests are also valuable. It is easy, then, for your support team to refer your question or request to the right person. And if things go poorly, you have evidence of having made a clear and reasonable request. (It is almost always unnecessary [and unwise] to "prove" that someone did not perform—but the fact that you do have documentation will be an incentive for that person to do something for you, just in case that person is not moved by the inherent value of your request.)

Remember, there are personalities involved in these transactions as well. One "workaround" solution is to substitute another member of your team for the one who lacks rapport with someone whose help your team needs. Suppose your operations engineer does not get along with the manager of the shipping department. It probably would be wise, then, for you personally to arrange for shipping. It would be nice to fix all "personality" problems, but since you are driven to make your project work in the here and now, you may just have to set aside the "improvement" of the work interface, and/or refer the problem to their FMs.

COMMUNICATION WITH THE CUSTOMER

In most companies, the PM is the primary interface to the customer. While your contracts administrator (CA) may handle formal correspondence, and your system engineering lead may do the technical heavy lifting, you remain responsible for the success of all these communications. As with the discussions above, you should remain vigilant for anomalies in the communications of all the members of your team with all the members of the customer's team. You should make sure that your customer-counterpart is comfortable discussing performance and communication issues with you.

Because customer interface issues are so vital to the success of your program, you should consider taking strong action if and when necessary. If your CA and the customer's buyer get into trivial arguments that derail progress, for example, in negotiations, then it may be necessary to change CAs. It might not be his or her fault, but that really does not matter to your program. You need to insist on good relations. Even if it turns out that you and your customer counterpart are unable to work together, it may be necessary to make a change. Depending on how important either of you are to the success of the program, a decision may need to be made between your manager and his or her manager. But it is way better to resolve any personality issues early and not let any of these issues get in the way of your ability to interface with the customer. Looking at the big picture, it may be more reasonable to replace the customer's PM, but remember, they are the customer and, therefore, have the most say in any of these conflicts. All the more reason to strive to work well with the customer.

A CRITICAL COMMUNICATION SKILL

If you do not get any tangible help from this chapter other than the following communication tips, you will have been well served. "Active listening," popular among counselors and parents, is a great communication tool. One way to ensure that you have correctly received a message that a person is trying to send to you is to restate what you heard in your own words. This obviously can become silly if overused:

> *John:* I'm going to get a cup of coffee, and I'll be right back.
> *You:* I see that you are planning to take a break and it won't be long before you return.

But for more complex messages, even technical ones, restating what you think you heard is very valuable. In transactions between colleagues and customers, where mutual understanding is vital, active listening is a terrific tool.

Sometimes, you may think you got the message, but when you restate it, it turns out that the message you heard was not the message that was intended to be sent. If the conversation were to then proceed with a shared misunderstanding, it might take quite a while to unravel the thoughts when the error is eventually detected:

> But I thought you said 5 dB, not 5 octaves!

If you had said, "that's dBs, right?" when you first heard the "5," you would have saved time and frustration.

Active listening is very helpful when you are trying to diagnose "people issues."

> "I think you're saying you don't like to work with John at all."
> "No, John is OK, but every time we have a small revision to make, he goes off the deep end."
> "Ah, I see. I'm glad that 99% of the time you work together well. Maybe you should talk to John about this quirk before the next small revision is needed."

If you think such an approach sounds more like psychology than program management, you have not grasped the wholeness of program management as both an art and a science. You must constantly be looking for ways that your program can get stuck and constantly diagnosing problems, including communications problems. If you have a scientific background, use that scientific curiosity for the good of the program as you investigate the <u>why</u> of communication problems.

SUMMARY

Poor communications are often blamed when programs go awry. But that diagnosis is useless if you do not take action, and the correct action, to make sure that "poor communications" does not adversely affect your program. We looked at who has to communicate with whom on your program and discussed the "why" of that poor communication trap. We reminded you that every problem on your program is your problem, and that good communication among your team is worthy of your constant attention. And last but not least, we reminded you of a simple communication device to build understanding and to develop even deeper meanings in complex communications.

END OF CHAPTER QUESTIONS

For Discussion

1. What kind of program environment is conducive to good communications? What can the PM do to build this communications culture?
2. Might people not tell the PM that there is trouble brewing if they think the PM will yell at them when they tell him or her? If this is not the fear, what reason(s) might be in play?
3. Two people on your program do not get along. You have assessed the depth of their distaste for one another and feel it would be unwise to get them together to discuss the issues. What can you do indirectly to help them progress in their working relationships?

Written Assignments

1. Describe the downward spiral of communication for unapproachable leaders.
2. Try as you might, you cannot get along with the manager of the shipping department. Now you need a "favor" to get expedited shipping on one of your products. What can you do?
3. Give an example of how active listening can help your program.

11

Earned Value Management

In earlier chapters, we discussed the value and use of earned value management (EVM). EVM is the most frequently used tool to make the management (actually, the *measurement*) of program progress more quantitative.

Years ago, a simplistic program management technique was to track spending against planned spending. If you were spending on plan, it was assumed that you were making the necessary progress. In conjunction with monitoring spending, the program manager (PM) would monitor schedules. If you were on schedule, it was assumed that your costs would be on schedule. Effective PMs would do both, of course—make sure that spending was on track and make sure that schedule progress was as planned.

EVM puts these two analyses together. EVM starts with a detailed schedule with clearly defined tasks. Those tasks are integrated with the estimated effort for each of those tasks. On many projects, those tasks are structured as part of a work breakdown structure (WBS). The estimated effort for each of those tasks typically was defined during the proposal phase of the program.

APPLYING EVM THEORY

So now you have a list of all the tasks that need to be performed, when they are supposed to be done, and how much each of those tasks is expected to cost. (As we mentioned earlier, it is a very good practice to attach a name to each task—be it an individual on your team or the leader of a subproject.) Figure 11.1 shows a simple illustration of this type of plan.

This illustration shows the steps necessary to design PC Board A, which is WBS 7.1 on your schedule. Notice that each task has been ascribed a

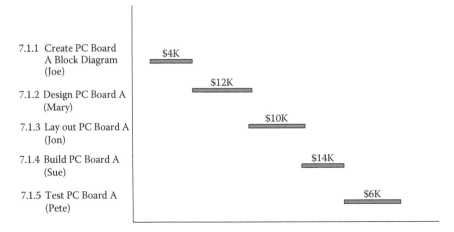

FIGURE 11.1

Earned value management (EVM) planning schedule.

planned value (PV). These costs are derived from multiplying each team member's hourly costing rate times the number of hours allocated to the task. (Note this is one place where some cost pluses or minuses can occur. When the program was estimated, it is likely that an average costing rate was used for each discipline. But in most accounting systems, the individual's actual costing rate is used. Thus, if the individual is highly paid versus the average, your costs will be higher than were estimated!) The PV for the entire task is thus $4K + $12K + $10K + $14K + $6K = $46K.

Note that some tasks can begin before the predecessor task is complete. As Joe is creating the block diagram, Mary can start selecting parts. As Mary is completing the design, Jon can start his part models for the PC board layout. While Sue cannot build the board until the design is done, parts can be ordered. And Pete can be setting up the test area for when the board is built. This overlapping of tasks, which is valuable for compressing the schedule and saving costs, does make the EV (earned value) analysis a little less "clean." Depending on your company's usual processes, the nature of the schedule complexity and task structure, and your (the PM's) preferences, you may or may not take partial credit for tasks that are in process. This estimate of partial completion is naturally arbitrary and can thus be misleading. That is why some practitioners use only completed tasks in their analysis.

In most companies, EV is calculated on a monthly basis. Figure 11.2 shows the schedule filled in for actual performance up to Point A, a

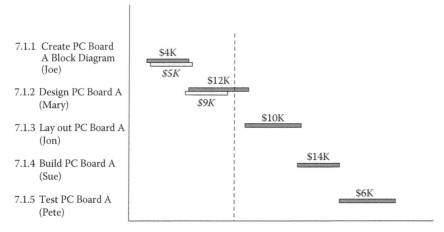

7.1.1	Create PC Board A Block Diagram (Joe)
7.1.2	Design PC Board A (Mary)
7.1.3	Lay out PC Board A (Jon)
7.1.4	Build PC Board A (Sue)
7.1.5	Test PC Board A (Pete)

FIGURE 11.2
EVM progress at Point A.

monthly program review, for example. We will illustrate the various EV parameters by using this example:

> PV is essentially the same as budget at completion (BAC). The primary difference being that your BAC may include some decrement ascribed to management reserve (MR). To keep this example simple, however, we will assume they are equal (i.e., MR = 0).

To this point, the PV is the completion of the PC board design (4K) plus about 80% of the design (12K × 0.8 = 9.6K) for a total PV of 13.6K.

> AC—Actual cost

Since both of these tasks are complete, the AC is simply 5K + 9K = 14K.

> EV—Earned value

The EV is the value of the completed work without regard to what the ACs were. In this case, the EV is 4K + 12K or 16K.

With these data, we can calculate one of the most important EVM values, CPI.

> CPI—Cost performance index

The CPI is the ratio of what the work actually costs divided by what the work was planned to cost. In this case:

$$CPI = EV/AC = 16/14 = 1.14$$

So clearly, a CPI greater than 1 is good—it means your work that was planned for a certain number of dollars was completed for fewer dollars.

EVM is also used to measure performance against schedule. By comparing the amount of work that was actually done to the work that was planned to be done, you can calculate the schedule performance index (SPI).

$$SPI = EV/PV$$

In this case, the first two tasks are done (EV = 16) and by the schedule task 1 (4K) and 80% of task 2 was supposed to be done (12K × 0.8 = 9.6K).

$$SPI = 16/9.6 = 1.67$$

Great! Your SPI is greater than 1, and thus you have accomplished more in this time period than you had planned.

Putting it back into context, this means that the first two tasks went very well. In fact, you have built a little reserve in that it cost you less to complete the work than you planned. This is a positive cost variance (CV):

$$CV = EV - AC = (4+12) - (5+9) = 2K$$

And you have a positive schedule variance (SV) also:

$$SV = EV - PV = 16 - (4 + [12 \times 0.80]) = 16 - 13.6 = 2.4$$

Note that your 2K of CV is real savings that can be applied to any other negative cost events that may occur. The positive SV is also good, of course, but depending on other factors, it may not help your overall schedule. For example, if Jon is not ready to start the PC board until the planned time, your SV is lost. Or if this WBS element is not on the critical path, there may not be a net time savings.

Returning to the big picture, work proceeds, and your next sample point is due. Looking at Figure 11.3, the next time EV is calculated, we are at Point B.

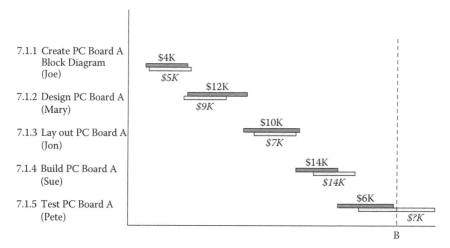

FIGURE 11.3
EVM progress at Point B.

The PV is:

$$4+12+10+14+6=46$$

The EV (since the PC board test—task 5—is not complete) is:

$$4+12+10+14+X$$

where X is the EV of the last task.

So for the analysis at Point B, X has to be estimated. Since task 5 is not complete, it is difficult to determine what the value should be. Talking to the engineer will give you some insight, perhaps, but in troubleshooting, until you get to the "Ah-ha" moment, you really do not know how long it will be. However, for purpose of the EV calculations, we will make an assumption:

Assume task 5 is 90% complete (after all, you have spent just about all of the PV of 6K at this point).

Calculating:

$$EV = 4+12+10+14(6\times0.9)=45.4K$$

$$AC = 5+9+7+14+(\text{actual expenditures on task 5}=4K)=39K$$

So CPI is:

$$CPI = EV/AC = 45.4K/39K = 1.16$$

which looks positive. You are almost finished with WBS 7.1, and you have a positive CPI. Let us see how the SPI is doing:

$$SPI = EV/PV = 45.4/46 = 0.987$$

So your SPI is less than unity, which means you are late; that, of course, is clear from looking at the schedule. What you do not know, though, is how late you are going to be. The SPI cannot help with predicting the future. In fact, a program may have a positive SPI and yet have a schedule completion problem brewing. How could this be? It is simply because the SPI measures the conglomerate schedule performance, but does not analyze for the critical path in your schedule. Your *critical path* is that set of tasks that strung out as predecessors and successors that form the longest interdependent set of tasks in your schedule. Thus, you can see that early completion of tasks that do not directly affect the completion date of the program, even if showing excellent SPI numbers, are completion-irrelevant unless they are on the critical path.

At this point, you might want to calculate where you are on costs for this program. You can take your CV and see how much you are ahead:

$$CV = EV - AC = 45.4 - 39 = 6.4K$$

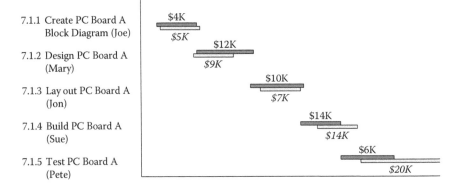

FIGURE 11.4
EVM—at completion of WBS 7.1.

Thus, you have 6.4K in the bank, and if you can finish task 5 for the nominal 6K plus the residual 6.4K, or 12K, you will be fine from a cost perspective.

The next step is to see what happens when that board is finally completed. Looking at Figure 11.4, note that task 5 took 20K to complete—which is more than three times the amount planned. This will no doubt adversely affect your cost and schedule.

$$CPI = EV/AC = (4+12+10+14+6)/(5+9+7+14+20) = 46K/55K = 0.84$$

Since it is less than 1, there is a problem—not good.
And how about the SPI:

$$SPI = EV/PV = 1$$

which indicates you are finished! (But not when you planned.)

Because your CPI is less than 1, you could expect a negative CV:

$$CV = EV - AC = 46 - 55 = -9K$$

This means it took you \$9K more to complete this WBS element than planned.

Now consider what your thought process would be if you were running this program. At the beginning, at the first sample point, Point A, you would be feeling pretty confident. If things continue to go this way, then you would expect to finish early and to underrun this task. But by sample Point B, things are beginning to not look as good. Although the CPI is still good, the SPI being less than 1 indicates the trouble that is brewing. But even then, it was close to 1, so you may not have been too alarmed.

USING EV INTELLIGENTLY

EV theory includes forecasting techniques that use existing performance to predict where the project will actually end. This is the TCPI (to complete performance index). TCPI is calculated as follows:

$$TCPI = (PV - EV)/(PV - AC)$$

You can see that if your ACs are less than your EV, TCPI will be greater than 1. If you can adjust your TCPI to 1, then you will meet your PV or budgeted costs. Of course, this means that your AC and EV are equal. But, it does tell you just how much better your AC has to be over your EV to make your cost target.

As far as predicting is concerned, basically all EV analysis can do for you is use past performance to predict future performance. (Since this is often misleading, investment offerings are required to say that "Past performance is not an indicator of future performance,"—as they are telling you in the same breath what the past performance has been.)

So, though generally not advisable, you can calculate what you expect your final value will be by the following formula:

$$EAC = PV/CPI$$

That is the estimate at completion (EAC), which can be calculated by taking the current CPI and dividing it into the PV.

This is exactly like saying, "I expect things to keep going as they are going." When you think about it, this is a simplistic analysis. Your program is made up of many different types of tasks done by various people. Really, the only common thread is perhaps that the tasks were all quoted by (or overseen by) a single person. So if the greatest factor in your CV is that the job was over-optimistically bid across the board, there might be some validity in this calculation. But, in general, predictions of the future are not helpful in that they can be more misleading than insightful.

There is nothing as effective in determining actual progress as conversations with the people responsible for the work. One way to formalize this conversation is to create a new EAC. In doing so, you ask each of the responsible managers for each active and upcoming WBS element to predict how much more money they will need to finish their tasks. If they say they need more time and money than planned, it is time to put your leadership hat on. Are they being overly cautious or are they alerting you to a serious problem? It is important to know the difference, and take the appropriate corrective action, which might include reminding them of their obligation to find a way to meet the original plan, releasing some of your reserves, or economizing in other areas. But, as always, open communication, in a fear-free environment, is the key to maximizing the probability of a successful completion of the program.

At the end, the CPI is clearly negative, but it does not do you any good to know it then. This example was chosen not only to explain the use of EV analysis but also to show that it *can* be misleading. The fact that task 5 took much longer than expected was the problem, of course. The extension of task 5 over what was planned can be due to any of three things or a combination of all three:

1. Your task 5 estimate was incorrectly optimistic.
2. Pete's performance was not good.
3. Things beyond your knowledge or Pete's control happened.
   a. For example, a marginal op amp was supplied on the board, and it was in a circuit location that was hard to find

Generally, when this kind of thing happens, folks tend to guess it is (2). Actually, in this author's experience, this is the least likely cause.

But, the other lesson is that simply watching the numbers closely does not predict how things will be performed, nor does it tell you what the problems are. An effective PM will use the numbers for some guidance, but will gather the bulk of his or her information by talking to the people responsible for the work as the program progresses. In our example, before the program was evaluating EV at Point B, the PM should have noted that task 5 started late. And a quick conversation with Pete would indicate how things were likely to occur. Pete would have the best estimate of how long it might take to complete task 5. More importantly, by talking to Pete, the PM might be able to help—does Pete need some technical consultation, a better oscilloscope? The numbers do not tell you what is needed, but by talking to Pete to collect your information, you are also able to learn things that might allow you to effect an improvement.

SUMMARY

EVM is a valuable tool, and is in widespread use, so you must understand it. But with that understanding comes the need to understand its limitations. The more predictable the work you are monitoring, the better the tool is—for example, for the 10th production run of some avionic equipment. But for design and development, the dependence on the correct estimating of a task during the proposal cycle makes supreme reliance on EVM somewhat risky. Especially for design and development programs, there is no substitute for frequent and frank discussions with those people doing the work. And always keep in mind that EVM is an indicator, not a control.

END OF CHAPTER QUESTIONS

For Discussion

1. Discuss the risks of using EVM as your primary method of measuring progress. How can these risks be mitigated?
2. Some people (the naïve ones) may think that EVM techniques control costs. But you are smarter, so describe some actual controls that you may use based on EVM analysis.

Written Assignments

1. Write a brief description, in your own words, of how EVM works.
2. How important is a good schedule, with realistic estimates of task content, to EVM? Did I hear you say "Very"—but now tell us why!
3. What are the best and worst kinds of contracts for the application of EV methods? And why?

Mini Project

1. This chapter uses a simple project to examine EV concepts. Develop one of your own, and if possible, use an example germane to your company's products and processes.

12

Negotiations

Of course, book after book has been written about how to negotiate, so it may seem a little ambitious to "cover" this large a topic in just one chapter. But the negotiations you will perform, lead, or participate in as a program manager (PM) are a controlled subset in the broad field of "negotiations."

One significant difference between general negotiations and those that a PM might engage is the context of those negotiations. If you are buying a car, relationships are very unimportant. You do not typically have a relationship with the car dealer, nor do either you or the dealer expect to form one. Sure, he or she would like you to come back when you want another car, but you both know that is not very likely in the near future. Thus, there is no past relationship framing the basis of your negotiation, and there is no need to maintain a relationship in the future. Therefore, you are both out to get the best deal that you can—"win–lose" is the operative engagement mode.

As a PM, the opposite is usually true. Most of your negotiations will be done in the context of relationships you already have or ones that you are currently building. And most of them will not be with customers. In fact, your day will be filled with negotiations of various kinds, and mostly they will be with people you work with on a frequent, if not daily, basis. Many simple work transactions are actually negotiations. They are also opportunities to make strategic as well as tactical progress. When you think about what is good for both you and for the other person, you have the opportunity to not only get what you want or need for the program, but you have the opportunity to deepen the trust and relationship that are the capital of the PM.

With whom would you negotiate, and why call it a "negotiation" anyway? Some of the following examples are clearly negotiations, but some transactions are negotiations de facto.

CONTRACT NEGOTIATIONS

A common situation is that your company is in a sole source situation on a product, and you have submitted a proposal price that your customer thinks is too high. Most of your negotiations with the customer will be based on Federal Acquisition Regulation. In these negotiations, ethical behavior is expected and even governed by law. Your estimates of the work must be current, complete, and accurate. You may not overstate an effort just to negotiate down to a lesser number, nor can your customer purposely understate an effort in order to get your price down. You cannot lie, although it may be not uncommon to use some partial truths. Of course, all estimates are opinions as to how much work needs to be done, so a difference of opinion can be expected. The better you have done your homework, and the better you feel about the accuracy and ethicality of your bid, the better the negotiations will go for you. You usually have one great advantage—you and your team have spent a lot of effort developing your proposal. Typically, your customer has spent only a very small percentage of that effort developing his or her expected value. Since disclosure of methodology is usually mandated, tell your customer how you developed your price and ask him or her to do the same. You usually do not have to state the obvious, but you may wish to say something in the most tactful way possible such as: "If we spent 300 hours preparing our estimate and you spent only a few hours, we feel our approach may yield more accurate results." Remember, there is an underlying requirement that both you and your customer adhere to—that the estimates are fair and in good faith. Since this is true, a discrepancy between your estimate and the customers can only exist because one of you saw the work incorrectly. If you had a team of people working on it and your customer had one (him or herself), logic indicates that your estimate would be better.

Your customer has an advantage too. Typically, he or she will lead the negotiations, and the format is also largely under the customer's control. One way to work toward an agreement is to look at the various parts and to discuss the content as you both see it. Suppose there are eight of these subgroups, and both you and your customer have each estimated all eight. The negotiations might proceed along the following lines:

- Task 1—You and your customer discuss how the estimate was made and note the discrepancy.
- Task 2—Same as task 1.
- Task 3—The customer says, "That one is pretty much in line, we can skip it."
- Task 4—Same as task 3.
- Task 5—Same as task 1.
- Task 6—Same as task 3 ("It's OK").
- Task 7—Same as task 3.
- Task 8—Same as task 1.

The customer then suggests that you work harder to resolve the discrepancies on tasks 1, 2, 5, and 8. That seems reasonable on the surface. The next step in your customer's approach is to resolve those discrepancies, and when that is done, negotiations are complete. Well, do you see what happened? The customer does not want to talk about tasks 3, 4, 6, and 7 *because your estimate was less than theirs!* If you meet halfway only on the tasks for which the customer wants to negotiate, you have been outmaneuvered. You should directly ask to discuss how much the customer has estimated for those other tasks. You could state your willingness, then, to accept his or her estimates of those tasks, just as you are being asked to accept the estimates that are lower. As you negotiate, remember that the direction the customer takes is the direction that is beneficial to his or her argument. Do not hesitate to open the discussion to things that the customer appears to accept as is.

If it is a pure price negotiation, resolution may be very difficult. Fortunately, there may be other parameters to with which to negotiate. Some requirements may be expensive, but not really that important to the customer. Removing requirements, and removing the costs for those requirements, can be a big step toward agreement. Also, making schedule adjustments that fit your company's staffing profiles are other ways that the customer can help reduce your costs, and therefore, your price.

The broader the negotiation field the more likely you are to find things that are important to you and not to your customer and vice versa. Looking for these things in a great collaborative way is a powerful path to a win–win solution.

When you are negotiating, always think win–win. If you are embarking on a multiyear contract with this customer, you both want it to be as positive as possible from the beginning. Understanding each other's situation is the key. In general, being open with your customer is the best way to have him or her to be open with you. Open, honest discussions are the best way to get to win–win solutions.

CUSTOMER NEGOTIATIONS—ONGOING CONTRACTS

When you reach agreement on price and scope and the contract is placed, formal negotiations may be over, but informal and frequent negotiations are just beginning. As you execute the contract, there will be things that you want (faster turnaround on approvals, for example) and things your customer wants (a copy of a related drawing that you are able to provide). My advice is to just "give" the customer anything that you properly can that is not too difficult or costly to give. Avoid a tit-for-tat approach—"If you do this, I'll do that." What's more effective for you is to help the customer any way you can and expect the same in return. It almost always works. You can enhance the probability by telling the customer that your plan is to do all that you can to be as cooperative as possible and to make him or her as satisfied as possible. If you and your customer are both interested in helping each other be successful, you both have significantly increased the probability of a successful relationship, and thus, a successful contract.

As with the initial negotiation, the goal should be to satisfy both your needs and the customer's needs. The more resourceful and creative you are, the more likely you can be mutually successful in these almost daily "negotiations."

INTERNAL NEGOTIATIONS—WORK BUDGETS

Your first (and it will not be your last) opportunity to negotiate internally will likely be with functional managers (FMs) as you develop your proposals. In general, your proposals will actually be based on estimates from FMs for the work they will be expected to perform under the contract. Since they will be expected to perform the work for the estimated value, there is a great incentive to estimate high. This is not unusual, nor is it really improper. They know that things can go wrong, and they know better than to estimate too aggressively. You, on the other hand, have a reason not to take the estimates carte blanche from the FMs. If everyone has a little "fat" in their estimate then your proposal will go in "a little fat." If it is fat enough to make all your FMs comfortable, you will not get the work. So it is important to challenge estimates that you may think are unnecessarily comfortable.

The more they expect that their feet will be held to the fire during execution, the more likely they are to estimate high. Your negotiations with them, therefore, have to paint a successful (and true) picture of how you expect to work together during contract execution. If they expect you to rigidly "enforce" their allocation, they will naturally be conservative when they estimate. To get them to estimate properly aggressively, they have to feel secure and that you will work with them through the natural bumps that will be encountered in any program. Here, the best credentials you have are how you treated overrun issues in the past. If you embarrassed or rebuked the FM, you can expect a tough negotiation in which win–win would not possible. On the other hand, if the last time the mechanical FM was overrunning, and you helped her out and figured a way to get the costing in bounds without allowing it to become an issue, she will not be as threatened by agreeing to aggressive goals. It will be a trusting working relationship, which is the best path to success.

So just as with the customer, there was a formal initial negotiation, and then a series of informal ones, so too it will be with your FMs. And these relationships are very important, since they will be challenged on a daily basis as you work with the FMs to make progress despite all the challenges that Fate has in store.

Putting the negotiation with the FM in context, after the contract negotiation described above, it is very likely that you have had to compromise on your price (and hour content) to some degree. In some cases, the agreement is directly connected to a particular task, but in many cases, it is unallocated. You may lose 5% of your design hours, for example. Each of the FMs is expecting (somewhat naively, I think) that they will be allocated all that they estimated. Most seasoned estimators, however, know there will be challenges from negotiations and challenges that are due to the need for you to create a management reserve. In most cases you will want to have (or be required to have) a reserve fund, linked to program risks, that can be used to cover cost "surprises." You must, however, be very frugal with releasing these reserves; if not, they will be gone before you know it! If the program was not quoted with risk dollars, you would likely create this reserve fund by allocating a little less than they quoted to the FMs to do the work.

This will be another opportunity to practice win–win negotiations— the FMs may want every nickel they quoted, and may not feel the reserve (which is potentially available to every function if needed) is really in everyone's best interests. As in your negotiations with the customer, you and the FMs want this to be win–win, since you will be working together for years.

Experienced PMs and FMs both know that the probability of things going exactly as planned is virtually zero. Scope issues will come up, both plus and minus. Work that was quoted by the system engineering manager may end up being done by one of the electrical engineers (EEs). You and they would waste valuable time making minor adjustments to the allocations. The best situation is when the FM is striving to get the work done for the lowest possible cost and at the highest possible efficiency, and you know this and trust him or her to do just that. Although the threat is that you will say, perhaps in a program review, "Mary (the EE manager) is over budget and needs to get her act together." No skillful PM would ever say that, of course. And Mary should know that you intend to help her meet her numbers and not wait to catch her not meeting them. If the environment is demanding, but not punitive, negotiating more ambitious goals with Mary will be much easier than if she feels threatened. Thus, in relationship-based negotiations, a feeling of trust within a supportive environment is your very best ally.

SUPPORT GROUPS—NEGOTIATIONS WITH SUPPORT GROUPS

Your program, most likely, is not self-contained. It is likely that you share a financial analyst, a contracts administrator, a financial analyst, and a scheduler with other organizations. On occasion, you will probably also need IT technicians, pack and ship services, security, reproduction, drafting, etc. These folks will have loyalties to their FMs and to the company but not especially toward your program. With good planning and no surprises, you will be able to give these people in their respective organizations plenty of time to do their jobs. However, I have never known a program to have such a combination of competence and luck that everything runs smoothly. Therefore, it is very likely that you will have to ask some of these colleagues to be especially attentive to your program and provide expedited responses. How you do this and what your track record and reputation are will greatly affect how much help you get. Meeting the folks halfway, or more, is a great way to get help: "Hey Bill, if I bring the work statement authorizations downstairs to you now, would you be able to sign them today?" (Now, if the last time Bill asked you to do something for him quickly you were "too busy," you can bet that Bill would be too busy for you also.) You never know when you will need someone's help, and it always pays to be cooperative and friendly. I have seen the "bluster" or "demand" approach, and I have seen that it does not work. People are people, and if you treat them well, they will help you. And you will always need help of one sort or another.

SUPPLIER NEGOTIATIONS

Most of the time, the procurement department will do the negotiations with suppliers, unless a technical or schedule issue arises. This is because in Department of Defense procurements, there are tremendous regulations to ensure equity in the procurement of materials. Because your buyers know these rules, and you do not, you must rely on them to do the bulk of the negotiations. Engineering managers or PMs may participate in setting price goals, but like your own negotiations with your customer, there are other parameters that need to be considered. Primary, of course, is delivery. Can you accept a partial delivery? Can you waive source inspection? Can you provide technical support for the first time build? Similar to all of your negotiations, practice win–win. You are not powerful enough not to want friends rather than enemies. No one is!

SUBCONTRACTOR NEGOTIATIONS

Many organizations have separate subcontract management departments, but depending on what you subcontract, you are likely to have a strong role in this process. If it is a question of doing the work in house or subcontracting it, or if it is a case of covering a peak load in engineering by bringing on an engineer-subcontractor, you may have internal negotiations as well. There may be a FM who does not want you to subcontract mechanical engineering work, for valid or invalid reasons, that you may or may not agree with given your program's requirements. The discussion that ensues is a negotiation. Again, you will need to maintain a good relationship with this FM, so listen to his or her concerns and resolve the issue. Sure, you may be able to pressure them into what you want, but what you really want is big-picture success. This FM may be a contributor or a detractor to that success. Of course, being agreeable does not mean that you never get your way—but it is <u>always</u> a mistake to appeal to your power to feed your ego.

Once the internal negotiations are concluded and it is decided that you will subcontract that mechanical engineering work, you will be part of the team that brings in the subcontractor. Where the work is performed, what the compensation per hour will be, and what the scope of work will be are all likely to be at least partially in your control and/or require your agreement and signature. So that signature is an obvious manifestation of your power—but again, use that power very sparingly and very wisely. Power not only corrupts, but the display of it makes more enmity than respect!

SUMMARY

Since almost all of your negotiations are with people with whom you will be working closely or at least peripherally, and because you are absolutely likely to need the cooperation and the help of these people to make your program a success, always enter negotiations with a win–win objective. Tell the person with whom you are negotiating that it is your intent to make the solution acceptable to both of you. Depending on the parameters that you can include in the negotiation and depending on how creative you are, you can increase the probability of a successful negotiation. While we did not discuss folks who do not know how to negotiate in a win–win way, most people, even if inexperienced in this type of negotiation, will come around quickly. But there are some folks who will want to trick you or will want to win at your expense. In that situation, I would recommend finding a way to work with someone else—perhaps their manager or colleague. Even if they attempt to trick you, and even if you are sure you can outsmart them, refrain from doing so. Outsmarting an adversary may feel good at the time, but the last thing you need is to have to work with someone you have outsmarted and possibly embarrassed. <u>Always</u> think win–win, and refuse to be diverted from that goal.

END OF CHAPTER QUESTIONS

For Discussion

1. In a contract negotiation, how can you capitalize on the hard work that you and your team have done in estimating the work?
2. Depending on your customer, it may or may not be a good idea to negotiate "tit-for-tat" but rather to help each other out wherever you can. Explain.
3. In "carving up the pie" (allocating budgets after negotiations to the functional departments), what environment will make the FMs likely to take on challenges? How do you create that environment?
4. Discuss if "being nice" to folks who provide services to your program (such as IT) is a sign of your weakness or strength. Would you be better off, in some cases, by being "demanding?"
5. Needing your signature on a subcontract agreement is a clear indication of your power. Describe some situations where use of that power is positive and other situations where it has more negative than positive consequences.

Written Assignments

1. Discuss how important existing and future relationships are with respect to the following situations:
   a. You are buying a car.
   b. Your customer and you disagree on scope on a long-term contract.
   c. The electrical engineering functional manager wants to "borrow" one of your program's engineers.

    d. Your contracts administrator feels you are being too generous in your interpretation of the specifications in your customer's favor.

    e. Your administrative assistant thinks it is not his job to retype the unformatted system engineer's report.

2. Describe a "trick," intentional or not, that your customer may use if negotiating eight distinct tasks that he or she has estimated as well. What can you do about it?

13

Coaching

Part of every program manager's (PM's) job, and, in fact, that of every leader, is to get the best performance out of people they work with regularly. Far from being PM-selfish, eliciting the best performance from the people on your program is your obligation to them, as well as your obligation to the program. Surely, the primary obligation of the PM is to the people on his or her own program. But take a broader view, and include those folks that *affect* your program. And take a still broader view and include all the folks that can affect the success of your organization.

RECOGNIZING INFLUENCE IN COACHING

It stands to reason that successful PMs quickly develop a great deal of leadership experience and thus, influence; and influence, not power, is the capital of the PM. PMs interact with people on their programs, with senior management, with junior engineer, with assemblers, technicians, administrative assistants, functional managers (FMs), and most importantly, customers. There is no other discipline in an organization that interacts and influences so widely.

You therefore might expect that successful PMs have the formula on getting people to do things, and in fact they do.

Nurturing that skill in others is part of the leadership obligation, and it is very important for the success of your program. If the electrical engineering (EE) FM disheartens the EEs working on your program, it is you (and of course, they) that suffer. Coaching peers or even more senior managers takes a great deal of finesse, lest you make an enemy. People do not like to be told they are doing things wrong, or that they can improve.

This is a great example from my personal life when my attempt at coaching was unsuccessful. My son was in an accelerated math class in high school and also in the honors chorus. He had to miss math class one day because of a performance. He responsibly got his homework from a peer and did the written homework. However, the other student forgot to tell my son that there were three axioms to memorize. When the class met the next day, my son got a zero on the quiz on those three axioms. My wife, who has motherly protective instincts (one of the most powerful forces on earth!), a great sense of justice, and a research-oriented mind, searched the student handbook and found out that if you were out for n days, you had n days to make up the work. Aha, my son should have had a full day before he had to take the quiz! (Don't worry; I'm getting to the coaching part of the story.) My wife set up a meeting for both her and me with "the Formidable Miss Jones" (which is how she had introduced herself at open school night some weeks before). Well, as luck would have it, my wife had a scheduling conflict, and I left to face the Formidable Miss Jones alone.

(See Figure 13.1 for my concept of the Formidable Miss Jones.) We spent a few minutes going over my son's grades, which were excellent (except for the one zero in question). The meeting was tense but cordial. I then introduced the concept that my son's zero should be eliminated from his grades because he did not have the prescribed (per the student handbook) one day to make up the work. (After all, as an engineer, reading and conforming to requirements is second nature to me.) Miss Jones (that is, the Formidable Miss Jones) did not see it that way. She said, "These honors students are always going off for one activity or another. Math is important, and it should not take second place to other activities." Although I appreciated her passion for her subject, I said, "Well, I agree math is important, but the handbook rule should prevail in this case." Miss Jones said, "Well, I've been teaching this way for 30 years, and I don't know why I should change now." As luck would have it, we were just studying the concept of continuous improvement at work, and looking at how we were traditionally doing things and how we could make changes and thus make improvements. Fueled by my continuous-improvement zeal, I told the Formidable Miss Jones "Well, if you've been doing it the same way for 30 years, don't you think it is time to make an improvement!?!" Surprisingly enough, she did not take the constructive suggestion positively. In fact, she leapt to her feet (surprisingly fast for someone who had been teaching calculus for 30 years!), and I recognized that our discussion had concluded.

So in this coaching attempt, I was a dismal failure. I think, in retrospect, I could have done a lot better—so as it turned out, it was a teaching moment *for me*! I realized in the tense situation, with no mutual rapport or respect, the chance of my making a successful coaching comment was zero (just like my son's grade!). You need to have a receptive audience if you are going to coach successfully.

Just to complete the story—my son kept his zero, but it really did not make much difference. He got an A for the course anyway.

After the "encounter," I thought I should mention how poorly it went to the head of the math department, a nice, smart, open-minded guy. I told him about the discussion with Miss Jones and asked if he would like to weigh in on this very important matter, which we all knew to be symbolic rather than really important. He said, "Oh no, we [the math department] are all afraid of Miss Jones." Now that was wisdom on his part. Another teaching moment for me! (Too late!) And when I got home, I was sure to thank my wife for sending me to the mathematical front lines without so much as a helmet for protection.

FIGURE 13.1
The Formidable Miss Jones (as I saw her).

(From Dollar Photo Club, File # 63315716 tsuneomp.)

DETERMINING WHEN TO COACH

Certainly many people recognize that they can improve in many ways and are open to coaching, at least in their minds—but maybe not as much in their hearts. Or at least, maybe they think they will ask for help when they need it. Of course, that situation opens up a great teaching moment, but often the need for help is some sort of tactical problem. You may see its root cause as a leadership deficiency, but the person needs help and may not see it that way at all. Skills that are more mechanical and non-interpersonal are less problematic for people to seek help on. In learning a new software package, for example, many people will seek help from their peers who have already used it. But even then it is surprising how many people will spend too much time trying to figure it out themselves. Here, a little well-placed humility from the PM is a great example: "If Charlie isn't embarrassed to ask for help, maybe asking for help is OK." Occasionally, folks will ask, "How do you think I should deal with Mike's chronic tardiness?" or "Should we tell Liz that she needs to write more concisely?" But very often the opportunity comes when people say something like—"We should tell Liz to write more concisely." This kind of suggestion is a great opportunity to open up a coaching discussion, as in, "Yes, good idea, but how do you think we should approach it?"

Depending on Liz's personality, if Irving walked up to her out of the blue and told her, "You should write more concisely," she may hear, "I think your writing is poor." When people's feelings get hurt, the healing process can take years, and during that period, and if Irving and Liz have to work together, your program suffers. In general, if you do not have confidence that Irving will do the coaching well, offer to do the "dirty work" yourself. Irving will be relieved and may even be interested in how you will do it. A teaching moment!

Subtle coaching such as the example above is one of the best tools you have to increase performance. One of the best subtle coaching techniques, of course, is a good example. But this can go somewhat unnoticed. You can increase the effectiveness of good example by mentioning what you *did not* do: "I considered complaining about the system engineering [SE] report in the program review, in front of the SE manager, Tom. But I thought better of it and instead went to see Tom in his office. I figured he would be less defensive that way; no one needs another enemy!"

Another extremely effective technique is to tell someone how much you admire a particular trait in someone else. "You know, Bob, I never hear Brian say a negative thing about anyone. If he sees a problem, he handles it directly and sensitively. That's probably one reason that he is such an effective FM." Bob hears two things:

1. You appreciate nonnegativity.
2. You think that addressing problems directly and sensitively is effective.

Another benefit of this approach is that in the process, you have given Brian some extra influence points with Bob. Now Bob sees Brian as effective, and sees him as effective because of his behavior. Increasing the influence of your effective colleagues makes them even more effective. This coaching is best done in general—not after Bob has said something negative about a colleague. You need to prevent defensiveness when you are coaching. If you come on too strong, defenses go up and the message is not appreciated.

Of course, subtle coaching by reference to another's good behaviors works only if you have Bob's respect in the first place. If he thinks your opinion is worthless, your coaching is meaningless.

Here's an example of my being coached and how it worked. I was a relatively young PM when one of my colleagues told me, "You know what's wrong with you, Charlie? You're too nice." I think I managed to say "thank you," because, after all, I am "too nice." I'm sure my colleague meant the comment as constructive, but the manner in which it was delivered left some room for improvement: "You know what's wrong with you?" is a pretty aggressive introduction. And the actual message, being "too nice," was and is contrary to my motivational belief system. People always use motivational techniques on others that are effective on them. I'm "too nice" because when someone asks me for something nicely, I am much more willing to do it. And, you know what, that is true for the great majority of people you work with. My colleague's comment was much more old school—intimidation as a management tool is now recognized as absolutely ineffective in the long run in increasing performance. It is now essentially passé, thank goodness. More "traditional" organizations are working on removing intimidation from the sociology of their cultures. Not to get too psychological, intimidation as an approach in child rearing (yelling, spanking, demeaning) is also (thankfully) out of vogue. People raised by parents who don't intimidate, are not naturally responsive to intimidation. Just demographically speaking, intimidation, which never caused people to think harder anyway, is phasing out as a growing percentage of the workforce was not raised in intimidation-control.

Thus, the "coaching" ("you're too nice") was too aggressive and was incorrect in content. There are two strikes, but the third strike may be more meaningful for this example: because I had little respect for my colleague's motivational skill, I paid it no heed. No coaching occurred; I remained "too nice." In fact, I took the criticism of being "too nice" as a compliment—criticism from people whose judgment you don't respect can actually be viewed as affirmation!

SUMMARY

Getting the best performance out of the people on your program is a part of the PM's job function. Your ability to share your leadership skills with people more junior or less experienced than you are in coaching leadership skills, is important for your program and for the good of the organization as well (since some of the people you help are shared with other programs). Your ability to successfully influence improved performance is based on the respect that your colleagues have for you, and the way in which you provide the guidance. Subtle coaching methods are more effective because they are less threatening and decouple the defensive–emotional factor from the intellectual content of the message.

END OF CHAPTER QUESTIONS

For Discussion

1. Many situations are aided by the application of a little humor. What could go wrong in using humor in a coaching situation?
2. What role does your *authority* play in your coaching? What role does your *influence* play in your coaching?
3. Can you delegate a coaching opportunity to another member of your team (for example, to the lead software engineer)? What considerations should you have if you think you might want to do so?

Written Assignments

1. Give some examples from your own life when you have been coached. What made it effective or ineffective?
2. Rewrite the following sentence to make it an effective coaching description:

   > When coaching a member of your team, it is always best to be direct—after all, we are all adults here, and people like to be corrected when they are wrong.

   Well, that was easy; now elaborate on why you rewrote it that way!

14

Inheriting a Program Already in Progress

In previous chapters, we have spent a great deal of time discussing how to propose programs, how to get them off to a good start, how to properly staff them, and how to form relations that will be important for the program's (and the program manager's [PM's]) success.

This is all fine in theory, but very often you will not have that luxury. Big programs may span several years, and over those several years, things happen to the PM who started the job. He or she may get promoted, may retire, may leave the company, or (gasp) may be removed because things are not going well.

When that happens, of course, it will be necessary to install a replacement PM. And, you may be that replacement. Thus, all of the team selection, all the financial planning, all of the customer relations are all in place, but have been centered on your predecessor. Depending on the circumstances of his or her departure and how much respect the team had for him or her, you will have a small or large challenge in being accepted by the team as their new leader. Ideally, the team would have a choice in the matter, but 99% of the time that is not practical.

You may know some of the people on the team, and that will be helpful—that is, if you have good relationships with those people. (Another example of how you want to make more friends than enemies!) You may also know the customer from previous work or from related programs. But maybe not.

BECOMING A MEMBER AND LEADER OF THE TEAM

Realize this new opportunity as a great time to practice your humility! The people on the program know way more about it than you do. Depending on the circumstances, the previous PM may be available to you for "training" purposes. It would be very helpful if you had a couple of weeks of overlap with him or her. Then, the previous PM could introduce you to the rest of the team and give you his or her vision as to how he or she planned to make the program successful. You will have to study the documents of the program (schedule, contract, work breakdown structure [WBS], financial analyses, etc.), but you will learn more and faster by talking to the other leaders on the program—functional managers (FMs), project engineers, etc.—depending on how your company is organized. In many companies, tasks that the PM is responsible for are shared with the program's lieutenants. You may want to leave those assignments as is—for example, on a production program, there may be an operations project manager (OPM) who keeps the production schedule details and updates the schedule. If your background is primarily in design programs, it is very important to have someone to do that for you. On the other hand, if that is the role that you came from, you will need less help there, but maybe more help from your financial analyst.

The keys to "working your way in" are demonstrating honest humility and directly asking for help. Make sure that any previous poor practices are changed, but make sure the changes are seen as positive. For example, if some meetings excluded peripheral folks who may or may not have something to add or who would benefit from the discussion, talk to those folks and see if they want to be included. Inclusion is always pretty benign, as long as the folks you include want to be included.

Of course, there is a natural resistance to change, and your new ideas may not be readily embraced. Communication is the key. If you want to make a change, explain why and ask for counter reasons. Inclusion of your team in making an important decision is tremendously important. One leadership foible is to a make a unilateral decision, and claim inclusion simply by telling the people why the decision was made. Surely it is better to explain the reason, but it is infinitely better to consult the affected people on the team and see if they embrace the idea, or can refine it. It may be good leadership (not bad) to be talked out of a change that the team thinks is a waste of time or is ineffective. Find out why it is not liked and who does not like it. Important decisions deserve the participation of the team. As you establish yourself as the new leader, your desire to include the ideas of the people on the team will give you quick acceptance.

If things have been running smoothly, your new team would want you to pretty much leave things alone and carry on as it was. If things were going poorly, then they (and your senior managers) may want you to make changes for the better. Find out quickly which of these thoughts is in the team's and your own manager's heads!

IMPORTANCE OF CONTINUITY

One mistake that people make at times like this is to change some of the leadership in favor of folks that they like and respect. On the surface, this makes a great deal of sense. Why would you want to entrust important roles on your program to people you do not know? Roy was your OPM on your last job, and you and he worked together terrifically. Deb, the current OPM, seems a little more disorganized than Roy. Roy is available, and it would be great to work with him again.

There are three good reasons not to change:

1. Deb knows the program and you do not.
2. You do not know Deb—you should give her a chance.
3. Changing her out in favor of Roy looks like cronyism—which it probably would be.

Deb is part of the team and replacing her on a whim not only will upset her, but it will also upset the people she works with on the program. And other people will feel threatened: "Hmm, they took Deb off, maybe the new PM had a favorite contracts rep too!" When deciding on the changes that you might want to make, think like a physician: first, do no harm.

Building relationships, learning the program, learning the customer, and keeping momentum on the program are all important and do not have to be done sequentially, but the building of relationships has positive side effects in all these areas. Sit down and talk to the financial analyst on the program and ask him or her how the costs look and what you should worry about or pay particular attention to in order not to exceed the available budget. Ask the electrical engineer (EE) FM to tell you a little about the EEs on the program and what their strengths are. And talk to the EEs individually also. Remember again that they know much more about the program than you do, and every conversation will not only build a relationship but will also give you knowledge.

FRESH EAC

One formal step that you might want to consider on your new program is to do a fresh EAC (estimate at completion). The program no doubt has a current EAC, but it may not be fresh. Often it is required to update the EAC periodically; every six months or a year is common. But even if you are not due to update the EAC by your procedural requirements, it is a good time to ask for a new estimate from each of the budget centers or FMs on your team. Requiring people to stop and validate the plans for the rest of the program will cause some delay, but actually a little pause to check progress against the big picture plan is good for everyone.

Of course, asking for a new EAC is always a little like playing with fire. The FMs will now have a chance to ask for more money, and will probably be able to produce excellent reasons why their initial estimates were correct then, but <u>now</u> new circumstances will increase the number of hours needed to complete the work. To quote one of my senior managers from 20 years ago, "New EAC?!?!?. That's just a license to steal!!"

Here is another great opportunity to practice your negotiation skills. All estimates should be set to be challenging, and folks will not sign up to challenging objectives if they feel that if they miss it by a little they will be punished. You would rather have someone estimate 100 hours and come in at 105 rather than have him or her estimate 120 hours (to be safe) and come in at 115! Goals should be set to be challenging but achievable, and should be embraced rather than feared.

Your new EAC will in effect be a contract with your senior management. It will be <u>your</u> EAC, and you should embrace it. It should be challenging, not padded. It might be wise to tell your manager how aggressive it is, so that if you do go over by a couple of hours, it will be seen for good performance against a challenging goal.

SUMMARY

Becoming a PM on a program that is already in progress brings a collection of issues that would not be there if it had been your program from the start. The key process of taking over the program and its people and making them your own lies in open communication and in relationship building. Maintain successful staffing and plans to prevent mistakes you might make and to minimize the change's impact on the team. Establish rapport, not dominance, with the team and they will work with you to train you and to help the program be successful.

END OF CHAPTER QUESTIONS

For Discussion

1. You have just been assigned to take over a program that is in trouble. So much trouble, in fact, that the prior PM has been reassigned to counting paper clips in the supply room. Discuss some of the things that you should do on your first few days in the PM chair.
2. You inherit a program that seems to be going well, but you are not sure that the money remaining equals the work remaining. Explain how you would use earned value management and discussions with the folks on the program to tell you if there is a problem or not.

Written Assignments

1. On that program described in "For Discussion" question 1, you know the lead mechanical engineer from working with her on a previous program. You do not think she is a very good leader. Maybe you should replace her as soon as possible? What personal and team dynamics should you consider before taking action?
2. For the program of "For Discussion" question 1, what do you think the esprit de corps is for the folks on the program? Knowing that dispirited workers are way less productive than confident and happy workers, you know you must address this issue—so, how <u>would</u> you address it?

Summary

This book attempts to give some practical and useful insights into the science and art of program management in high-tech, complex environments. It attempts to weave theory and practical experience to be a resource to current program managers (PMs) and to those individuals who are beginning or considering a program management career. My own experience was that, as a young engineer, I considered program management a kind of superfluous position—the real nuts and bolts of a program were done by the engineers, of course! Only when I was forced (rather, perhaps, "encouraged") into program management did I appreciate the challenge and satisfaction that a PM would encounter. Since I genuinely like working with people, there is no more "working with" role than program management!

As a reminder, and as a convenience to those readers who would rather use this book for selected topics, I will provide a brief review of the content and main points of each chapter.

CHAPTER 1: INTRODUCTION

As a career, program management is not for everyone. An effective PM has talents and interests both in "management" and "leadership." PMs must be able to analyze data and determine what action to take to make their program technically, schedulary, and financially successful. That is the "what" of program management. But knowing what needs to be done is useless if the PM cannot get it done. This is where leadership, with its primary tool, influence, is needed.

The PM must extend his or her influence not only over people on the program, but on senior management, customers, and other "helpers," such as contracts and IT. The PM is part of a larger system of people with varying degrees of dedication and capability, and people who would may have other objectives that are not completely aligned with the success of the program. The PM must understand these parameters and plot a course for the program that capitalizes on the strengths and best motives of the people on and around the program.

A PM must establish a leadership role on the program. This requires the manifestation of important traits. There are certain "prerequisite" traits that any respected leader must have, including integrity, intelligence, and confidence. There are the "soft" traits that separate true leaders from true managers: compassion, collaboration, persuasiveness, and humility. And there are "hard" traits: forceful, analytical, and demanding.

One's skills can be developed, but talents are more in the fabric of the character or at least the personality of the individual. Effective PMs acquire the skills they need, and they recognize where their talents lie (or more importantly, where they are lacking). Building a program team, thus, becomes a process of forming alliances with people who have the talents in which the PM is deficient. Honesty and humility are thus the prerequisites of finding those colleagues—if you think you are great at everything, you would not need any help.

Programs come in various types. The primary types are research, design, and production. Contracts that govern these programs also come in various types: time and material, cost plus, and fixed price. While one might expect a close linkage between program type and contract type, circumstances and customers conspire to make this not straightforward. A PM must assess the risks and challenges of the program in the environment of the contract type, and use this understanding in plotting the successful course for the program.

The PM is described as the conductor of a symphony orchestra. Beautiful music will be made only through the leadership and technical competence of conductor. Just as a conductor coordinates the musicians in the orchestra, a PM coordinates the work of the people on the program, in collaboration with the functional departments.

Success for the program and the people assigned to it rests squarely on the PM and his or her ability to understand and lead the systems—contractual, functional, technical, and human—that comprise the program.

CHAPTER 2: LEARNING THE ROPES: UNDERSTANDING THE CULTURE, THE CUSTOMER, AND THE PROGRAM CAPABILITIES

PMs recognize that their programs exist in three overlapping environments: company, customer, and team. A systems engineering view of the composite environment takes into account the interaction among these three environments and their corresponding "cultures."

The company environment will influence team behavior. Communication is a prime example. A culture that is retribution-oriented will reduce communication comfort and may lead to withholding important information that the PM needs. If the PM recognizes this, however, he or she can build a local environment where all news, good and bad, is welcome and appreciated. Company procedures that govern program activities and determine team members' roles and responsibilities can be both helpful and hurtful to the program. The PM must draw the best from the company infrastructure and limit the negative effects, such as narrowing of perceived responsibility and overreliance on metrics.

The customer and his or her culture is the next major environment for the PM to "manage" through. The customer has three major responsibilities to the management of the contract: cost, schedule, and quality. The customer is as interested as the PM in the success of the program and thus can be an ally when things get rough. Building rapport and trust with the customer is one of the most important obligations of the PM.

The team is the engine of the program. The PM's leadership of the team is key to its success. Leadership traits such as trust, confidence, and humility are the tools that will build good team performance. In general, building ownership of the program among the team members is more important than a design choice, and effective PMs must subjugate their egos to the greater good of successful leadership dynamics.

An effective PM understands the three environments of the program and recognizes the limited control that he or she may have on some of those environments. But by understanding the influences and by understanding the people involved and their individual goals, local environments and important relationships can be built that draw strength from these environments and mitigate any negative effects.

CHAPTER 3: IDENTIFYING OPPORTUNITIES

As a company decides what new business to pursue, the PM in a related area may be one of the most knowledgeable about the risks and opportunities of the new venture candidate. These opportunities may be "strategic"—involving growing into a new, possibly adjacent business area, or more "tactical," involving, for example, expanding capabilities of a system in current design or production. The knowledge and insights that the PM has are useless if he or she does not have the influence to make his or her viewpoint known and appreciated.

The PM's knowledge and experience will help the company evaluate:

- The "realness" of the opportunity—is the customer serious?
- The risks—both financial and reputational.
- The necessary commitment—is the opportunity worth the resources that would have to be directed toward it
- The competition—is there a "hungry" competitor? Is someone on the "inside" already?

Just as the PM's knowledge and insight are applied to opportunities for the company, an effective PM is wise to weigh opportunities for himself or herself just as judiciously. It is unwise for a PM to accept an assignment that does not align with his or her talents or does not allow a reasonable expectation of success. The success of the company depends on the success of its programs and the success of its programs depends on the success of its PMs.

CHAPTER 4: PRE-PROPOSAL WORK

Programs are often won or lost before the request for proposal (RFP) is released by a customer. A PM (who might be acting as the proposal manager at this phase) should influence the decision as to how much time and money should be spent on preparing for the emerging opportunity. If strategically important, the company might want to invest in necessary technology to increase the probability of winning an important contract. To manage proposal expense and identify opportunities that are meaningful, effort may be expended to learn more about the opportunity and the competition's capability and interest. (Of course, this must be done ethically, despite the temptation to collect data from improper sources.) And in some cases, it might seem unnecessary or unwise to invest in pre-proposal work at all—one could just bid to the spec and hope for the best. This approach has some severe risks—bid too low and you might win a program at a cost you cannot achieve; bid too high and you may lose the respect of the customer.

For the best possibility of a successful program, pre-proposal work is invaluable. Understanding the risks of the proposal and guiding senior decision makers based on that knowledge can have a strategic effect on the company. By blending knowledge of the customer, the technology, and the capabilities of the company, the PM/proposal manager is in an excellent position to guide the company to the right opportunities and to guide it away from programs that everyone wishes had never been won.

CHAPTER 5: THE PROPOSAL PROCESS FOR A "TYPICAL PROGRAM"

While no program is "typical," the process for proposal development has some common characteristics regardless of the proposal type. All proposals should begin before the RFP is received. In almost all cases, either because the opportunity is with an existing customer or because the opportunity is one that has been researched and targeted, much is known, and can be known, before the RFP is received. Every legal effort to understand the customer's expectations and interests, as well as possible competitors' weaknesses and strengths, needs to be evaluated. Effort spent in this period is very valuable and is high-equity. The "bid"–"no-bid" decision is obviously made during this pre-proposal stage, and considerations such as probability of winning, desirability of the work, performance risks, strategic value, resource availability, and potential profit levels are considered. Very often, the future PM will be acting as the proposal manager. The proposal manager needs to believe in the value of the opportunity, and, therefore, is usually a strong advocate to pursue the opportunity. Challenges to this desire are to be expected, as cautious senior managers and colleagues may find reasons to avoid the risks and recommend no-bid.

Assuming a bid is authorized, the proposal manager must establish his or her win strategy, keeping in mind that this should be a developing and consensus-based process, even as the bid decision is being made; that is to say, one should not bid a contract unless one knows how to develop a winning proposal. The better the customer's needs and desires are known, the better the strategy that can be formulated.

The concept of "price to win" (PTW) can be a valuable tool in the formation and review of the proposal, but it is necessary to understand that the accuracy and precision of the PTW is limited and it must be used with caution, lest incorrect inferences are made.

Depending on the nature of the proposal, it may be prudent, if not necessary, to include exceptions or clarifications with the proposal. These clarify your offer and prevent the inclusion of what you might consider to be unnecessary costs. Of course, any time you tell a customer you do not want to provide a feature or want to limit performance, you run the risk of alienating him or her and, thus, whenever possible, it is good to try to determine how important a given issue or feature is to your customer.

Keeping the proposal team moving in the same direction, incorporating the desired themes into various sections, and all the while meeting scheduled dates, is a big job. Storyboarding is an approach to coordinate content and insure your proposal vision continues to be uppermost in the proposal authors' minds. Ghosting the competition's weaknesses (indirectly saying how qualified your company is in areas where you think your competitors are weak) is an important proposal device.

Developing a correct and aggressive enough price is very important. Because you will be bringing together estimates from many functional departments, you must be careful that your team is not going to "pad" their numbers to make it comfortable for them should your company win the proposal. The more your company's culture is supportive (rather than punitive), the easier it will be for you to develop sufficiently aggressive pricing. Building a passion to win among the estimating managers is also a key process to getting to that winning price.

Once the proposal text and price are developed, the work is far from over. To make the proposal as good as possible, there are several reviews that are typically held. Each of these reviews carries with it the potential for rework, in fact potentially significant rework. Allowing time for this rework as you prepare the schedule is vital to getting a quality proposal submitted on time. After edits by a multiparty team, there are many possible sources of error and confusion, and the value of a final read-through is emphasized.

Noncompetitive proposals, wherein your company has already been selected for the work, can be easier to win, of course, but they carry with them other issues and considerations. The customer knows that you know that you are the sole source, and, thus, is more likely to scrutinize your estimates in the belief that you may not feel as driven to develop aggressive performance strategies.

With the proposal submitted, and if you are selected, you must then finalize the contract's price and content. These negotiations are key to establishing a collaborative yet respectful relationship with the customer. And should the proposal not win, the possibility of challenging the choice of a competitor is an important consideration. At a minimum, a losing proposal is an opportunity to learn from the experience and to better understand your own effort and how it was viewed. It is also an opportunity to gain insight into the customer's thinking to better prepare you for the next opportunity.

CHAPTER 6: PLANNING THE PROGRAM AND STARTING WORK

Once a program is won, the real work begins for the PM. It is combination of management (schedules, budgets) and leadership (team building and selection, building enthusiasm). Most of the management part is covered by your company's processes and procedures, while it provides less guidance on the "softer" side of the job—the leadership part. But even establishing the schedules and budgets, although very much based on the proposal, requires leadership skills as you negotiate with the people on your team and their functional management. You will need to tabulate and quantify risks, and set up reserve funds for those risks.

One of the key leadership tasks is the establishment of the program team. In most companies, this is not the sole province of the PM or the functional managers (FMs). It is necessary to collaborate with the FMs (and the team members themselves) to build the most effective team possible. Developing and continuing to emphasize the team's shared vision is one of the most powerful ways to insure that the team is pulling together toward the same concept of success. A shared vision can control overtime costs and merge your company's rules with the work ethic and personal availability of the team members. Building teams with "external" individuals or organizations has both benefits and drawbacks, and the effective PM weighs these carefully.

Staffing your program with talented, dedicated people is the most important part of your job in this phase—and keeping them coordinated and working together efficiently is next in importance. A program culture of mutual caring, including especially the PM, is the best way to insure people work together. A "blaming" culture causes defensive behavior and finger-pointing, and the PM's job is to make sure that each member of the team is committed to the team's results, not to an individual's ego or image.

CHAPTER 7: RUNNING THE PROGRAM

With the program planned and kicked off, the effective PM knows that plan will not last long before the first event that requires change. This is the stuff of leadership—maneuvering your program though the rough waters that will no doubt be encountered. Thus, "leadership style" is key to dealing with the issues that invariably will arise.

The most effective leadership style is the "we are all in this together" approach. Older styles, perhaps more effective in more "traditional cultures" really do not work in today's workforce environment. Leadership and management merge during the running of the program. Management tools, such as schedule tracking, earned value (EV) metrics, and program reviews must combine with leadership skills to convey the need to perform to the schedule to the entire team. As you use EV metrics, you must continually be aware that EV is an instrument, not a control, and like any instrument, it is subject to limitations and calibration. Be especially aware of indicators looking good while the team expresses worry!

There is a strong need to be ever vigilant to the quality of your team's work and to infusing the quality culture into your program. Quality is the root of customer satisfaction and the vehicle to continued business. And you must be sure that your customer knows of your commitment to quality. His or her perception of your quality may be almost as important as quality itself.

Customers play an important part of the system that is your "program." Customers' personalities, interest levels, and talents come in wide ranges, and it is your job to make sure that you establish the best possible relationships with all ranges. Dealing with difficult situations on scope or performance is made much easier if a strong relationship has been established in the "good times." Knowing when to escalate or de-escalate is the PM's job. Only he or she is likely to have the circumspection to know how to resolve scope issues.

Program traps, such as the desire for perfection and scope creep (both internally and externally generated) can derail your program as easily as a tough technical problem. The PM must reel in the tendency to constantly make things better. Another trap is "getting stuck" on an technical problem—knowing how and when to intervene requires a good understanding of the egos involved and the tendency of engineers to borrow in and try to solve all the problems of the world themselves.

Customer interactions with the team—reviews and demonstrations—can help the PM with the team's schedule commitment. No one wants to disappoint a customer and you can use this driving force on your team just as effectively (almost) as it works on you!

Cost control is another part of the PM's job that has both a management and a leadership component. Reviewing cost data and lists of who is charging your program and why—these are the "management" part. Knowing what to do about it and then doing it is the "leadership" part. Like many interactions with people outside your immediate team, it is an opportunity to make friends or enemies. Always chose the approach that makes friends.

Part of the PM's job is to interface with senior management and transform their concerns into support for the program and the team. Avoid the negative feeling that performing program reviews for senior management is a distraction from your duty. It is part of your duty. It is your prime opportunity to build confidence in the team and to get help when you need it.

One of the chief duties you have in running the program is to be vigilant to budding problems. And once detected, your challenge is to mitigate them effectively and early. Rather than relying on metrics, the best early warnings are learned by talking to your team. It is in the trenches, through mutual trust and respect, that you have the best opportunity to detect the problems and steer the direction of the program. If you wait for problems to show up on your EV "instruments," the recovery will be much more difficult. This is why caring for your team, as individuals, is important to effective program management. Caring for them includes realizing that extended hours over long periods is not only bad for them; it is bad for your program. And appreciating the impact and importance of family commitments is the path to building the respect and rapport that are the basis of your effectiveness.

Rapport and respect are also the bedrock upon which to weather the changes that occur on the program, both natural (as phases come and go) and "unnatural" (changes in scope, funding delays). And, of course, the biggest change: the end of the program. All the work leads to that successful conclusion—it can only be successful if it is successful for the customer, your company, and your team. This is program management success!

CHAPTER 8: CLAIM IDENTIFICATION, CLAIM MANAGEMENT, AND CLAIM AVOIDANCE

There are many ways your program can go awry, and many of them are (or should be) in your control. Customer behavior is one thing that may be out of your direct control. Remember that you and your customer share a common goal, and that realization and helping your customer to remember it is the best way to avoid situations that might result in a claim.

While sometimes there is no alternative to a formal claim, most of the time the negatives on both sides will outweigh the positives, so your primary goal is to avoid having to need one. Working through problems with your customer, and helping him or her to remember their obligations, is the best way to avoid claims. However, even so, you must perform your role of protecting your company. An important way to help both your customer and yourself avoid problems, but to be ready in case some do arise, is by preparing a monthly progress report. While it is a nonthreatening and possibly informal document, it can be a record of your progress and your attempts to avoid problems.

By use of the report, for example, you can provide early warnings of upcoming events, or provide early notice of defective government-furnished equipment. Fomenting issues or noncontractual issues can be semiofficially diffused, and concerns about the correctness of customer specifications can be brought up for resolution with less potential for over-reaction on either side.

While the goal is to mitigate problems before they get out of hand, it is necessary to be ready for the occasional situation that they might. Thus, it is important for the PM to keep the contracts department aware of these issues and seek their collaboration and advice.

Remember, too, that claims can go both ways. If you have major subcontractors working on your program, you need to be sure that you treat them as you would like to be treated by your own customer. Be vigilant toward what you owe them and establish a rapport with them as well. Fighting a claim from a subtier can be a major distraction for you and your team, so all the strategies you use to make sure things are going well with your customer should be used when you yourself are the customer.

Work with your customers and suppliers as you do with your team—the most powerful way to do this is to embrace the philosophy of "we are all in this together."

CHAPTER 9: LEADERSHIP MODELS

Humility, and the recognition that the PM does not have all the knowledge necessary to make all the decisions on the program, is the key to effective leadership. It is the PM's leadership skills, not the "power" of the office, that gets things done.

Leading well-intentioned people is not really that difficult. Using the "we are all in this together" philosophy, blended with the nearly universal desire of people to be successful, makes guiding your team to a goal relatively easy. Arrogant "leaders" can make this difficult—arrogance foments rebellion. It is much more effective if a decision is the team's decision, not the one dictated by the PM. It is even possible that a "better" decision held by the PM is less effective than a "lesser" decision owned by the team. It may take a good deal of self-discipline on the part of the effective PM to let his or her good idea be supplanted by one recommended by the team that seems less practical. It also takes some self-confidence to see that agreement with the team is a sign of leadership strength, not weakness.

Leaders with high integrity and concern for the people on the team will be effective leaders. Leaders are effective when they listen more than they talk and when they support rather than demean.

CHAPTER 10: COMMUNICATIONS

An underlying theme of effective program management is effective communications. Recognizing that a PM cannot help situations of which he or she is unaware, an effective PM thinks about communications all the time.

An important realization is that people speak when it seems to be to their advantage to do so. That is why PMs must welcome all news, bad as well as good, and act in the best interests of the program, and, hence, in the best interests of the news-giver. The best communicators are the most nonjudgmental and non-overreactive listeners.

Effective PMs also monitor communication lapses among the team. As much as one might want everyone to work together comfortably, issues between team members can get in the way. Some of these can be easily fixed and some cannot. The effective PM recognizes the difference and either helps to resolve the problem (in the former case) or rearranges assignments in the latter. The principle of having the team members select new members with whom they want to work is the ideal, but of course must be modulated based on reporting structures and availability.

The PM is also responsible for effective communication outside the team—with other common departments such as shipping and security. If one of these functions (or a person in it) is allowed to become hostile to your program, it can hurt you badly before you realize it. The culture of respectful communication that you foster among your team will carry to the people your folks deal with.

The PM is the chief communicator with the customer, and the importance of that communications channel being open and clear cannot be overstated. Good communications with the customer is the basis of collaboration with him or her, and that collaborative spirit can get the program through very difficult situations.

All effective communication involves mutual understanding of the message. The principles of active listening are very powerful to this. They can be used not only to acknowledge receipt of a message, but to take it one step further: "OK, I see that you think this idea won't work. Do you have another idea that might?"

CHAPTER 11: EARNED VALUE MANAGEMENT

Earned value management (EVM) is the quantitative way to put the two most important parameters of a program's progress, schedule, and spending, together for analysis. It is a very powerful tool used to *measure* a program's performance, but, of course, it can never control it. Measurement is passive, but control is active. EV is best used on programs that are well known and well scheduled, such as the fifth run of a production program. Using EVM on design programs can be effective also, but the challenge is to have a very accurate view of the work to be done and the time that it should take to do it.

Overreliance on EVM on design or ill-defined programs can cause problems to be missed, and it is recommended that EVM be heavily supported by communication with the people responsible for the work. EV parameters, be they favorable or unfavorable, are based on the schedule that is the basis for them. It is entirely possible that the team is doing a good job, but that the EV numbers are poor, because, for example, the schedule was overly optimistic. That doesn't mean that there is not a problem—because the schedule is the promise to the customer and it is the basis for planning costs and resources. But it is important not to improperly demean the performance of the team. Bad EV numbers, like any other problem on a program, are meant to be solved in the same way—by the team, with your leadership!

CHAPTER 12: NEGOTIATIONS

As a PM, most of your negotiations will be in the context of ongoing relationships, with customers or team members. There is therefore a strong imperative to make these negotiations successful for both parties. You would not want to work with a customer who feels you outsmarted or overpowered him or her in a "heated" negotiation. Rather, if you think "win–win" from the beginning, you have an excellent chance of a successful negotiation. The more creative you are at the table and the better you are at understanding what your customer needs (vs. wants) and the better you understand the external pressures on him or her, the more likely you are to negotiate to a mutually beneficial solution.

While win–win may be your goal, you must be prepared to deal with "opponents" who are "win–lose" oriented. In the ideal, you convert their thinking to one of mutual satisfaction, but you must be ready for some hardheads. The best defense is to be negotiating from a strong basis—your position is fair and well thought out, and you truly want to see the other's viewpoint. Understanding your counterpart is key. Again, listening is better than talking—you don't learn anything when you are talking, and the more you know, the better you will do.

The same principles apply to internal negotiations, with engineers or FMs. Like external negotiations, there may be things that are important to you that are not important to your counterpart. Good negotiators seek out these imbalances and can use them to trade off things they don't need for things they do. When these can be found, it's possible for both sides to "win"—how powerful is that?

Thinking and acting win–win, with a true desire to get to "fairness" is the way to make negotiations successful.

CHAPTER 13: COACHING

Coaching people that a PM works with is part of his or her strategic obligation to the program and to the company. Getting the best performance from everyone who is on the program (or who affects it) is the best way to succeed.

Everyone needs help sometimes, and everyone can improve. An effective PM models this in his or her own behavior, making it more palatable for members of the team to seek help and advice. Thus, the first rule of coaching is that the "coachee" has to be willing to be coached, and the second is that the coach's advice is seen as valuable. The more subtle this transaction is the better, since the more subtle it is the more comfortable it is for both parties. Recognizing and talking about good traits in others is one way to encourage those traits in other members of your team. If you admire someone's diligence, you show that diligence is an important trait to you, and, if you are respected by your team, you may just have raised the "diligence quotient" by a few percent in one brief discussion.

And be prepared to be coached yourself—ask your team how you can do a better job of leading them. It may be a little threatening and uncomfortable, but the benefits to your knowledge of yourself and your team make the risks very worthwhile.

CHAPTER 14: INHERITING A PROGRAM ALREADY IN PROGRESS

While it might be nice and holistic to propose a program, win it, negotiate it, win it, post-negotiate it, plan it and staff it, and run it to completion, it doesn't often happen that way. Life's variables get in the way. Thus, it is likely that you will "inherit" a program that is already underway.

You will have to assess for yourself how healthy it is—by reviewing data, by performing an EAC (estimate at completion), but more importantly, by talking to the people on the program. These discussions will tell you much more than the empirical data, and will help you form the relationships that will make you successful. First impressions are very important, so your philosophy (which should be team-oriented, and personally humble) should show forth. You are not coming to "take charge"—you are coming to contribute to the team—albeit in a leadership role.

If the program is in trouble, you may have to make changes in direction, roles, or even in the personnel. These should be done cautiously and collaboratively—the people on the program will need to feel safe and appreciated if they are going to work though this change as your ally rather than as your adversary. Your new role is new for you, but it is also new for the members of the team who may have had a range of relationships with your predecessor and may expect them to continue.

Your goal should be to work with the team to provide continuity of the good aspects of the program and to provide shared direction to solve the problems that are there. Make yourself available and open to the members of the team, treat them with respect and inquisitiveness, and you will be quickly accepted as their new leader!

FINAL THOUGHTS

Program Management is an important and fulfilling calling. It is not for the weak of heart, the egotist, or the unfeeling. True career satisfaction can be found by those leaders who are dedicated to their customer, their company, and their team.

Index

A

Active listening, 212
Actual cost (AC), 217, 219, 222
Aerospace companies, leadership in, 197
Autocratic leaders, 199

B

Bid decision process, 58–59
Bid specification, 51
Black Hat review, 70
Blame culture, 67–68
Brain-holistic approach, 2
Budgeted work (BW), 111
Building relationships, 243
Business growth, PM, 5
Business reviews, 72–73
BW, *see* Budgeted work

C

CA, *see* Contracts administrator
Career, program management as, 1–2
Caring for team, 151–152
Caring leaders, 155
Casual overtime (OT), 94–95
CFE, *see* Customer-furnished
         equipment
Change management, PM, 6
Claims
    issues, 174–175
    management, 191–192
    of subcontractor management,
         190–191
Coaching
    determining period for, 238–240
    recognizing influence in, 235–237
Collaboration traits, *see* Soft traits
Collaborative relationship, 15
Collective group, 199
Commitment for program manager, 42

Communication-failure diagnosis,
         205–206
Communications, 205–208, 242
    among the team, 208–210
    with customer, 211
    outside the team, 210–211
    verbal, 32
Communication tool, 212
Company culture
    program in, 27–31
    team interaction with, 36
Company–customer interface, 37
Compensated OT, 95
Competition, program manager, 42–43
Competitive bidding, 22
Contract engineers, 93
Contract letter, 117–118
Contract negotiations, 226–228
Contracts, 16, 18–23
    concept of, 185
    delayed approvals, 179–181
    final negotiations, 74–75
    refinement, 75–77
Contracts administrator (CA), 23, 181,
         185, 189
Control approach, 164–166
Cost control, 143–146
    program and customer, 31–33
    project managers, 5
Cost impact, 179
Cost performance index (CPI), 112–113,
         217–218, 220–221
    detect problem in, 128–131
Cost Plus Award Fee (CPAF), 19
Cost-plus contracts, 16, 18–23
Cost Plus Fixed Fee (CPFF), 18, 19
Cost Plus Incentive Fee (CPIF), 18–19
Cost-reimbursable contracts, 22
Cost-sharing concept, 20–21
Cost variance (CV), 218, 220–221
CPAF, *see* Cost Plus Award Fee

CPFF, *see* Cost Plus Fixed Fee
CPI, *see* Cost performance index
CPIF, *see* Cost Plus Incentive Fee
Critical communication skill, 212–213
Culture program, 100–101
Customer demonstration, 123
Customer engineers, 184
Customer-furnished equipment (CFE),
    174
Customer-induced problems, 188
Customer-inflicted problems, 175
Customer interface management, 4
Customer management, 117–119
Customer negotiations, 228
Customers
    approval action, 179
    approval rights, 181
    communications with, 211
    defective information, 187–190
    program and, 31–34
    team interfaces with, 37
CV, *see* Cost variance

D

Defective government-furnished
    equipment, 177
Defense companies, leadership in, 197
Department of Defense contracts, 69
Design programs, 16–17
Dictators', 199

E

EAC, *see* Estimate at completion
Earned value (EV), 28–29, 112–113,
    216–219
    limitations, 141–142
Earned value management (EVM) theory
    applying, 215–220
    using intelligently, 221–223
EE, *see* Electrical engineer
Effective program leadership, 195
Electrical engineer (EE), 34, 105
Engineering labor, 92
Enlightened Program Manager (EPM),
    106
EPM, *see* Enlightened Program Manager
Estimate at completion (EAC), 222, 244

EV, *see* Earned value
EVM, *see* Earned value management
Experience for program managers, 12
External scope creep, 138–140

F

Federal Acquisition Regulation (FAR), 41,
    69, 77
Final proofreading, importance of, 71
Firm Fixed Price (FFP), 19
Fixed-price contracts, 17, 19–20, 31
Fixed Price Incentive contracts, 20
Flawed technical specifications, 185–187
FMs, *see* Functional managers
Functional management, 5
Functional managers (FMs), 137–138,
    195–196
    negotiations, 228–230
    performance issues for, 160–164

G

Gather intelligence strategy, 48–51
GFE, *see* Government-furnished
    equipment
Good communication, 208
Government-furnished equipment (GFE),
    173, 174
    defective, 176–179
    late, 175–176
Green Team review, 69–72

H

Hard traits for program managers, 11
Hold-them-accountable style, 104
Humility, program manager quality, 199
Hunting party, 107

I

Inappropriate comments, 181–183
Infantry platoons, 198
Inheriting a program
    becoming a member and leader of the
        team, 241–243
    building of relationships, 243
    estimate at completion (EAC), 244

Instrumentation approach, 164–166
Intellectual property monitoring, PM, 6
Internal negotiations, 228–230
Internal research and development
     (IRAD), 47–48
Internal scope creep, 134–138
Investment strategy, 47–48
IRAD, *see* Internal research and
     development

K

Knowledge, program manager, 39–43

L

Labor costs, 18
Leadership, 7–8, 87–92, 151–152
   deficiencies, 238
   misunderstandings of, 198
Leadership models, 195–197
   achieving and maintaining respect,
      200–203
   leadership theory examination,
      198–200
Leadership styles, 104–108
Leadership theory examination, 198–200
Left-brain managers, 2
Level of effort (LOE) program, 179
LOE program, *see* Level of effort program
Lump sum contracts, 20

M

Management, 84–87
   interface of program manager, 4
Management reserve (MR), 84, 129
Management style, 107
Matrix management structure, 5
McGregor's theory, 198
Metrics, 111–114
Micromanagement, 35
MR, *see* Management reserve

N

Negotiations, 225–226
   contract, 226–228
   customer, 228

internal, 228–230
   subcontractor, 231–232
   supplier, 231
   with support groups, 230–231
Noncompetitive proposals, 73–74
Noncontractual direction, 183–185

O

Official negotiation level, 187
Ongoing contracts, 228
Operations project manager (OPM), 242
Opportunities, 38
   for program manager, 43–45
Opportunity register, 87
Organization, departmental interfaces,
      23–24
OT, *see* Overtime
Outsourcing of product, 98–100
Overrun scenario, FFP contracts, 20
Overtime (OT), 94–95

P

Participative leadership, 198–199
Peer critique, 35
Personal traits for program
      managers, 11
Piling on syndrome, 35
Planned value (PV), 216–217, 219
PMs, *see* Program managers
PO, *see* Purchase order
Poor communication, diagnosing, 209
"Power of the purse" concept, 196
Pre-proposal period, 57–58
Pre-proposal work
   efforts to develop winning proposal,
      51–53
   strategies of work and investment,
      47–51
Prerequisite traits for program managers,
      11
Price to win (PTW), 62
Pricing, 67–68
   strategies and risk management,
      68–69
Prior performance, 61
Procurement process, 58
Production programs, 17–18

Program, 103–104
    changes and continuity, 152–156
    in company culture, 27–31
    completion of, 166
    confronting defeats, 157–159
    cost control in trenches, 143–146
    cost of, 31–33
    cost performance index (CPI), 128–131
    customer management, 117–119
    customers as motivators, 123–125
    dealing with individual performance
        problems, 160–164
    diagnosing and resolving problems,
        164–166
    in DoD/high technology arena, 15–18
    external changes, 156–157
    focusing on quality, 114–117
    forces and priorities, 133–134
    leadership and caring, 151–152
    leadership styles, 104–108
    making and monitoring progress,
        108–111
    metrics, 111–114
    monitoring *vs.* controlling, 141–142
    overview, 167–169
    performance traps, identifying and
        avoiding, 119–121
    problem with engineer, 121–123
    quality of, 33–34
    reviews, *see* Program reviews
    scheduling of, 33
    scope creep, *see* Scope creep
    senior management, 125–127
    and team, 34–38
    team problem, 131–133
Program analyst, 13, 14
Program life-cycle manager, 7
Program management, 1–3, 24–25, *see
    also* Program managers
Program managers (PMs), 1–2, 24–25,
    81–83, 103
    and communication, 211
    contracts, types of, 18–23
    experience of, 12
    knowledge of, 39–43
    negotiations and, 225
    obligation of, 235
    opportunities, 43–45

organizational overview of,
        23–24
    program's purse strings control,
        196–197
    programs, types of, 15–18
    qualifications of, 11–12
    role of, 3–8
    skills of, 14–15
    talents of, 13
    understanding of strategic goals, 195
Program planning, 81–84
    documents, 96
    leadership part, 87–92
    management part, 84–87
    outsourcing product, 98–100
    outsourcing work packages, 97
    program culture, building, 100–101
    sourcing, 92–96
Program reviews
    monitoring schedules, 146–150
    senior management, 126–127
Progress monitoring, 108–111
    of program manager, 4
Proofreading, importance of final, 71
Proposal managers, 3, 22–23, 44
Proposal preparation—storyboarding,
        66–67
Proposal team, 63–64
PTW, *see* Price to win
Purchase order (PO), 74–77, 81, 181, 186
PV, *see* Planned value

Q

Qualifications of program managers,
        11–12
Quality control, program and customer,
        33–34
Quality of teamwork, focusing on,
        114–117

R

Red Team, 69–70
Relationship-based leadership style, 104
Request for equitable adjustment (RFEA),
        174, 190
Request for information (RFI), 53–54

Request for proposal (RFP), *see* Request
    for quotation (RFQ)
Request for quotation (RFQ), 41, 47, 57, 58
Research-oriented type of program, 15–16
Respect in leadership models, 200–203
RFEA, *see* Request for equitable
    adjustment
RFI, *see* Request for information
RFQ, *see* Request for quotation
Right-brain managers, 2
Risk management, 68–69
Risk register, 87
Risks for program manager, 41–42

S

Schedule control
    management, 5, 18
    program and customer, 33
Schedule impact, 179
Schedule performance index (SPI), 112,
    113, 218, 220, 221
Scheduler/planner, PM, 4
Schedule self-management, 108–111
Schedule variance (SV), 218
Scope creep, 140–141
    external, 138–140
    internal, 134–138
Secret technology, 74
Self-confidence, 8
Self-justification, 14
Semi-formal report, 117
Senior management, 62, 69, 75, 125–127
Senior manager, 107–108
"Shared goal" concept, 33
Sign-off process, 72–73
Skills of program managers, 14–15
Small-scale proposals, 72
Soft traits for program managers, 11
SOW, *see* Statement of work
Specification management, 6–7
SPI, *see* Schedule performance index
Spice model, 120
Staffing in project management, 5
Statement of work (SOW), 67
Storyboarding, proposal preparation,
    66–67
Strategic planning, PM, 5–6

Strategy development, 60–62
Subcontract management, 190–191
Subcontractor negotiations,
    231–232
Subtle coaching methods, 238, 239
Superior technology, 60
Supplier negotiations, 231
Support groups, negotiations with,
    230–231
SV, *see* Schedule variance

T

Talents of program managers, 13
TCPI, *see* To complete performance
    index
Team
    building, 64–65
    communications, 208–211
    dynamics, 66–67
    program and, 34–38
Test specification (T-spec), 115
Time and material (T&M) contracts, 16,
    21–22, 32, 42
Time-sensitive task, 110
T&M, *see* Time and material contracts
To complete performance index (TCPI),
    221
Top-level proposal schedule, 63
Troubleshooting, defective GFE, 177
T-spec, *see* Test specification
Typical program, proposal process for
    bid decision process, 58–59
    business reviews, 72–73
    contract refinement, 75–77
    Green Team review, 69–72
    noncompetitive proposals, 73–74
    pre-proposal period, 57–58
    pricing, 67–68
    pricing strategies and risk
        management, 68–69
    proposal preparation—storyboarding
        and team dynamics, 66–67
    proposal team, 63–64
    strategy development, 60–62
    team building, 64–65
    winning contract—final negotiations,
        74–75

U

Underrun scenario, FFP contracts, 21
Unenlightened Program Manager (UPM),
    105–106

V

Verbal communication, 32
Vicious compliance, 34
Vision statement, 89

W

WBS, *see* Work breakdown
    structure
Winning proposal development, pre-
    proposal efforts to, 51
Win strategy development, 60–62
Work breakdown structure (WBS), 63,
    84, 215
Work budgets, 228–230
Work environments, 196